Electronics: Technology Fundamentals

Electronics:
Technology Fundamentals

Christopher Gunn

Clanrye International
www.clanryeinternational.com

Clanrye International,
750 Third Avenue, 9th Floor,
New York, NY 10017, USA

ISBN: 978-1-63240-906-5

Cataloging-in-Publication Data

Electronics : technology fundamentals / Christopher Gunn.
p. cm.
Includes bibliographical references and index.
ISBN 978-1-63240-906-5
1. Electronics. 2. Electronic apparatus and appliances. I. Gunn, Christopher.
TK7835 .E44 2019
621.381--dc23

For information on all Clanrye International publications visit our website at www.clanryeinternational.com

CLANRYE INTERNATIONAL

Printed in China.

Contents

Preface

Electronics is a branch of physics and engineering. It is concerned with the development and application of devices and systems that involve the flow and control of electrons. The movement of electrons can occur in matter and vacuum or in semiconductors. Electrical circuits that are designed from different active and passive electrical components are studied in this field. Active components include transistors, diodes, etc. Passive components are resistors, inductors, capacitors, etc. Most electronic devices use semiconductors today. Some of the branches of electronics are analog electronics, digital electronics, microelectronics, optoelectronics, integrated circuits, etc. Circuits and components used in electronics can be analog or digital. This book aims to shed light on some of the unexplored aspects of electronics. Most of the topics introduced herein cover new techniques and applications of this field. Those in search of information to further their knowledge will be greatly assisted by this book.

Given below is the chapter wise description of the book:

Chapter 1- Electronics is the science concerned with the flow, emission and control of electrons in matter or vacuum. This chapter has been carefully written to provide an easy understanding of the various functional elements of an electronic circuit, such as electric power, inductance, capacitance resistance, voltage and electric current.

Chapter 2- The field of electronics has a number of significant branches, such as digital and analogue electronics, microelectronics, power electronics, optoelectronics, etc. This chapter closely examines the fundamental principles and concepts related to these fields for a holistic understanding of electronics.

Chapter 3- Electrical elements are the components of an electrical circuit. These may refer to active and passive electrical components. Some of the important active elements are diodes, transistors and batteries, while passive elements may include resistors, capacitors and inductors. This chapter discusses in detail the working of these electrical elements and their significance in an electrical circuit.

Chapter 4- An electronic circuit is a system of electronic components connected by electrical wires for the flow of electricity. The primary electronic circuits are AC and DC circuits. All the diverse aspects of AC and DC circuits, such as resistance, impedance, phasors, transfer functions, etc. have been extensively detailed in this chapter.

Chapter 5- Some of the important instruments used in electronics are ammeters, ohmmeters, multimeters, oscilloscopes, logic probes, spectrum analyzers and signal generators. This chapter analyzes their working principles and their use in the measurement of physical quantities and testing in electronics.

Chapter 6- Electronics is a vast subject that has a wide range of applications in areas such as audio electronics, radio electronics, nanoelectronics, molecular electronics and bioelectronics, among others. The diverse applications of electronics in the modern day have been thoroughly discussed in this chapter.

Indeed, my job was extremely crucial and challenging as I had to ensure that every chapter is informative and structured in a student-friendly manner. I am thankful for the support provided by my family and colleagues during the completion of this book.

Christopher Gunn

1

Understanding Electronics

Electronics is the science concerned with the flow, emission and control of electrons in matter or vacuum. This chapter has been carefully written to provide an easy understanding of the various functional elements of an electronic circuit, such as electric power, inductance, capacitance, resistance, voltage and electric current.

The word electronics is derived from electron mechanics, which means to study the behavior of an electron under different conditions of applied electric field.

The branch of engineering in which the flow and control of electrons in vacuum or semiconductor are studied is called electronics. Electronics can also be defined as the branch of engineering in which the electronic devices and their utilization are studied.

The motion of electrons through a conductor gives us electric current. This electric current can be produced with the help of batteries and generators.

The device which controls the flow of electrons is called electronic device. These devices are the main building blocks of electronic circuits.

Electronics have various branches include, digital electronics, analog electronics, micro electronics, nanoelectronics, optoelectronics, integrated circuit and semiconductor device.

Consumer Electronics

Consumer electronics are electronic equipment intended for everyday use by people. Consumer electronics usually find applications in entertainment, communications, and office productivity.

Some categories of consumer electronics include telephones, audio equipment, televisions, calculators, and playback and recording of video media such as DVD or VHS.

One overriding characteristic of all consumer electronic products is the trend of ever-falling prices. This is driven by gains in manufacturing efficiency and automation, coupled with improvements in semiconductor design. Semiconductor components benefit from Moore's Law, an observed principle which states that, for a given price, semiconductor functionality doubles every 18 months.

Many consumer electronics have planned obsolescence, resulting in E-waste.

Electronic Components

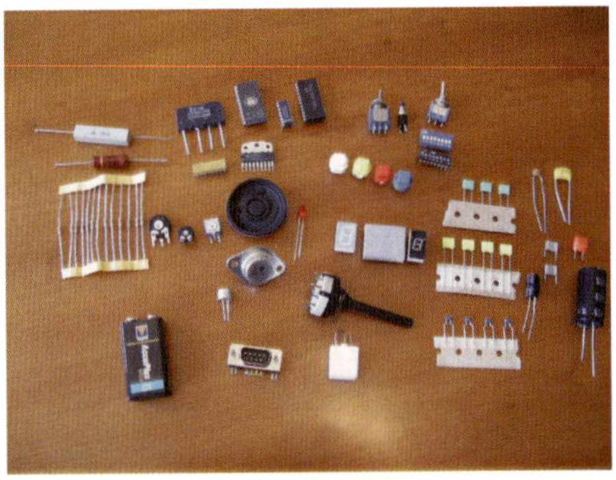

Figure: A Components of electronic devices.

An electronic component is a basic electronic building block usually packaged in a discrete form with two or more connecting leads or metallic pads. The components may be packaged singly (as in the case of a resistor, capacitor, transistor, or diode) or in complex groups as integrated circuits(as in the case of an operational amplifier, resistor array, or logic gate). Electronic components are often mechanically stabilized, improved in insulation properties and protected from environmental influence by being enclosed in synthetic resin.

Components are intended to be connected together, usually by soldering to a printed circuit board, to create an electronic circuit with a particular function, such as an amplifier, radio receiver, or oscillator.

Many electronic components are on the market today.

Types of Circuits

Analog Circuits

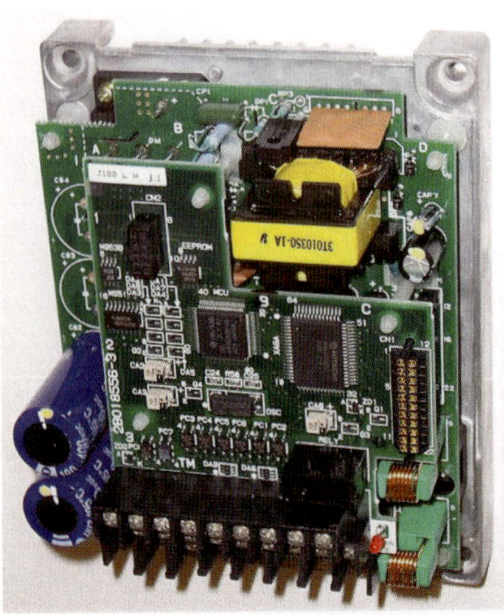

Figure: Analog Circuits.

Most analog electronic appliances, such as radio receivers, are constructed from combinations of a few types of basic circuits. Analog circuits use a continuous range of voltage as opposed to discrete levels as in digital circuits. The number of different analog circuits so far devised is huge, especially because a 'circuit' can be defined as anything from a single component, to systems containing thousands of components.

Analog circuits are sometimes called linear circuits although many non-linear effects are used in analog circuits such as mixers, modulators, etc. Good examples of analog circuits include vacuum tube and transistor amplifiers, operational amplifiers and oscillators.

Some analog circuitry these days may use digital or even microprocessor techniques to improve upon the basic performance of the circuit. This type of circuit is usually called 'mixed signal'.

Sometimes it may be difficult to differentiate between analog and digital circuits as they have elements of both linear and non-linear operation. An example is the comparator which takes in a continuous range of voltage but puts out only one of two levels as in a digital circuit. Similarly, an overdriven transistor amplifier can take on the characteristics of a controlled switch having essentially two levels of output.

Digital Circuits

Digital circuits are electric circuits based on a number of discrete voltage levels. Digital circuits are the most common physical representation of Boolean algebra and are the basis of all digital computers. To most engineers, the terms "digital circuit," "digital system" and "logic" are interchangeable in the context of digital circuits. In most cases the number of different states of a node is two, represented by two voltage levels labeled "Low" and "High." Often "Low" will be near zero volts and "High" will be at a higher level depending on the supply voltage in use.

Computers, electronic clocks, and programmable logic controllers (used to control industrial processes) are constructed of digital circuits. Digital Signal Processors are another example.

Building-blocks:

- logic gates
- Adders
- Binary Multipliers
- flip-flops
- counters
- registers
- multiplexers

- Schmitt triggers
- Highly integrated devices:
- microprocessors
- microcontrollers
- Application specific integrated circuit (ASIC)
- Digital signal processor (DSP)
- Field Programmable Gate Array (FPGA).

Mixed-signal Circuits

Mixed-signal circuits refers to integrated circuits (ICs) which have both analog circuits and digital circuits combined on a single semiconductor die or on the same circuit board. Mixed-signal circuits are becoming increasingly common. Mixed circuits contain both analog and digital components. Analog to digital converters and digital to analog converters are the primary examples. Other examples are transmission gates and buffers.

Heat Dissipation and Thermal Management

Heat generated by electronic circuitry must be dissipated to prevent immediate failure and improve long term reliability. Techniques for *heat dissipation* can include heatsinks and fans for air cooling, and other forms of computer cooling such as water cooling. These techniques use convection, conduction, and radiation of heat energy.

Noise

Noise is associated with all electronic circuits. Noise is generally defined as any unwanted signal that is not present at the input of a circuit. Noise is not the same as signal distortion caused by a circuit.

Electronics Theory

Mathematical methods are integral to the study of electronics. To become proficient in electronics it is also necessary to become proficient in the mathematics of circuit analysis.

Circuit analysis is the study of methods of solving generally linear systems for unknown variables such as the voltage at a certain node or the current though a certain branch of a network. A common analytical tool for this is the SPICE circuit simulator.

Also important to electronics is the study and understanding of electromagnetic field theory.

Electronic Test Equipment

Electronic test equipment is used to create stimulus signals and capture responses from electronic Devices Under Test (DUTs). In this way, the proper operation of the DUT can be proven or faults in the device can be traced and repaired.

Practical electronics engineering and assembly requires the use of many different kinds of electronic test equipment ranging from the very simple and inexpensive (such as a test light consisting of just a light bulb and a test lead) to extremely complex and sophisticated such as Automatic Test Equipment.

Computer Aided Design (CAD)

Today's electronics engineers have the ability to design circuits using premanufactured building blocks such as power supplies, resistors, capacitors, semiconductors (such as transistors), and integrated circuits. Electronic design automation software programs include schematic capture programs such as EWB (electronic work bench) or ORCAD or Eagle Layout Editor, used to make circuit diagrams and printed circuit board layouts.

Advantages of Electronics

Electronic devices are playing a major role in everyday life. The various electronic devices we use in everyday life include:

- Computers

Today, computers are using everywhere. At home, computers are used for playing games, watching movies, doing research, paying bills and reservation of tickets for railways and airlines. At school, students use computers to complete their assignments.

- Mobile phones

Mobile phones are used for variety of purposes such as for sending text messages, making voice calls, surfing internet, playing games, and listening songs.

- ATM

ATM is an electronic telecommunication device particularly used for withdrawing money at any time from anywhere. ATM stands for automated teller machine. The customer can withdraw money up to a certain limit during anytime of the day or night.

- Pen drive

Pen drive is particularly used for storing large amount of data and also used for transferring data from one device to another. For example, the data stored in the computer can be transferred to the pen drive. The data stored in this pen drive can be retrieved at anytime.

- Television

Television is an electronic device primarily used for entertainment and knowledge. It is used for watching movies for entertainment, news for knowledge, cartoons for children's.

- Digital camera

Digital camera is a camera used for taking pictures and videos. This images and videos are stored for later reproduction.

Voltage

Voltage, also called *electromotive force*, is a quantitative expression of the potential difference in charge between two points in an electrical field.

The greater the voltage, the greater the flow of electrical current (that is, the quantity of charge carriers that pass a fixed point per unit of time) through a conducting or semiconducting medium for a given resistance to the flow. Voltage is symbolized by an uppercase italic letter *V* or *E*. The standard unit is the volt, symbolized by a non-italic uppercase letter V. One volt will drive one coulomb (6.24×10^{18}) charge carriers, such as electrons, through a resistance of one ohm in one second.

Voltage can be direct or alternating. A direct voltage maintains the same polarity at all times. In an alternating voltage, the polarity reverses direction periodically. The number of complete cycles per second is the frequency, which is measured in hertz (one cycle per second), kilohertz, megahertz, gigahertz, or terahertz. An example of direct voltage is the potential difference between the terminals of an electrochemical cell. Alternating voltage exists between the terminals of a common utility outlet.

A voltage produces an electrostatic field, even if no charge carriers move (that is, no current flows). As the voltage increases between two points separated by a specific distance, the electrostatic field becomes more intense. As the separation increases between two points having a given voltage with respect to each other, the electrostatic flux density diminishes in the region between them.

Exmple

Take a regular 1.5V AA Battery for example, between it's two terminals (+ and -) there's a potential difference of 1.5 Volts.

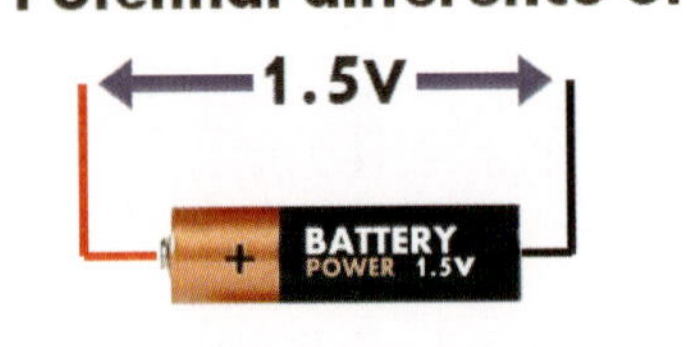

Voltage, or potential difference, is simply a measurement of the amount of energy (Joules) per unit of charge (Coulomb). E.g. in a 1.5V AA battery, every coulomb (charge) will receive 1.5 volts, or Joules of energy.

Voltage = Joules ÷ Coulomb

1 Volt = 1 Joule per coulomb

100 Volts = 100 Joules per coulomb

1 coulomb = 6,200,000,000,000,000,000 electrons (6.2×10^{18})

Voltage is always measured across two points with a device called a Voltmeter. Voltmeters are either analog of digital, the latter being the most accurate. Voltmeters are usually built into portable digital multimeter devices, as illustrated below, they are a common and often essential tool for any electrician or electrical engineer. You will usually find analog voltmeters on older electrical panels such as switchboards and generators, but almost all new equipment will come with digital meters as standard.

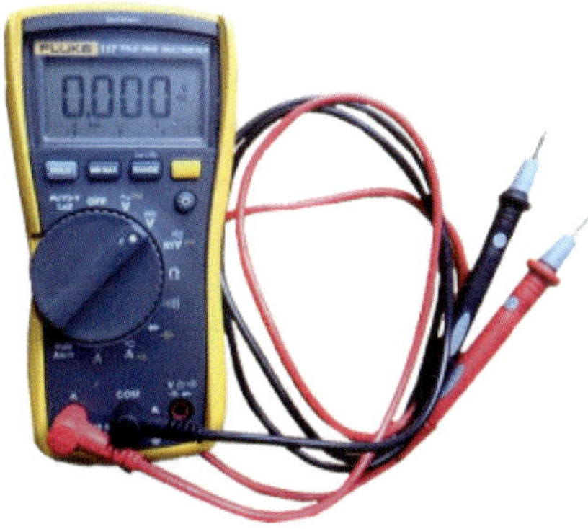

Portable digital multimeter with a voltmeter function

Analog Voltmeter

On electrical schematics, you will see voltmeter devices denoted by the letter V within a circle, as illustrated below:

Voltmeter symbol

Calculating Voltage

In electrical circuits, the voltage can be calculated according to Ohm's triangle. To find voltage (V) simply multiply current (I) by resistance (R).

Voltage = Current / Resistance

Example

Current in circuit = 10Amps

Resistance of circuit = 2Ω

Voltage = 10A x 2Ω

Answer: 26V

Electric Current

Electric current is defined as the rate of flow of negative charges of the conductor. In other words, the continuous flow of electrons in an electric circuit is called an electric current. The conducting material consists a large number of free electrons which move from one atom to the other at random.

Unit of Current

Since the charge is measured in coulombs and time in seconds, so the unit of electric current is coulomb/Sec (C/s) or amperes (A). The amperes is the SI unit of the conductor. The I is the symbolic representation of the current.

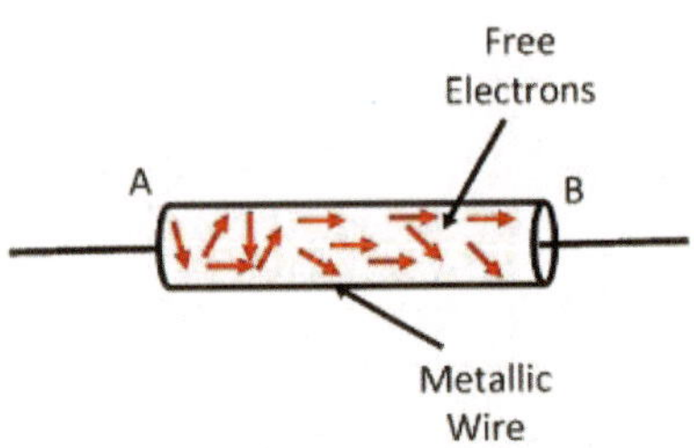

Q = 1 coloumb; t = 1 second; then I = 1A Thus, a wire is said to carry a current of one ampere when charge flows through it at the rate of one coulomb per second.

When an electrical potential difference is applied across the metallic wire, the loosely attached free electrons start moving towards the positive terminal of the cell shown in the figure below. This continuous flow of electrons constitutes the electrical current. The flow of currents in the wire is from the negative terminal of the cell to the positive terminal through the external circuit.

Conventional Direction of Flow of Current

According to the electron theory, when the potential difference is applied across the conductor some matter flows through the circuit which constitutes the electric current. It was considered that this matter flows from higher potential to lower potential, i.e. positive terminal to the negative terminal of the cell through the external circuit.

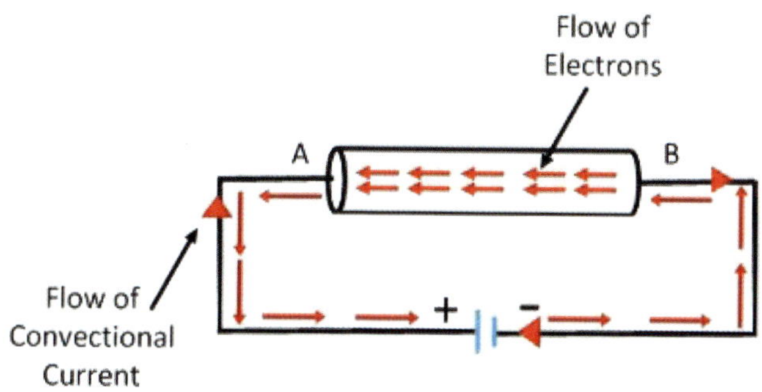

This convention of flow of current is so firmly established that it is still in use. Thus, the conventional direction of flow of current is from the positive terminal of the cell to the negative terminal of the cell through the external circuit. The magnitude of flow of current at any section of the conductor is the rate of flow of electrons i.e. charge flowing per second.

Mathematically, it is represented by

$$I = Q/t$$

On the basis of the flow of electric charge the current is mainly classified into two types, i.e. alternating current and direct current. In direct current, the charges flow through unidirectional whereas in alternating current the charges flows in both the direction.

Electric Power

The rate at which the work is being done in an electrical circuit is called an electric power. In other words, the electric power is defined as the rate of the transferred of energy. The electric power is produced by the generator and can also be supplied by the electrical batteries. It gives a low entropy form of energy which is carried over long distance and also it is converted into various other forms of energy like motion, heat energy, etc.

The electric power is divided into two types, i.e., the AC power and the DC power. The classification of the electric power depends on the nature of the current. The electric power is sold regarding joule which is the product of the power in kilowatts and the running time of the machinery in hours. The utility of power is measured by the electric meter which records the total energy consumed by the powered devices. The electric power is given by the equation shown below.

$$Electrical\ \ Power = \frac{Work\ done\ in\ an\ electrical\ current}{time}$$

$$P = \frac{V/t}{t} = VI = IR^2 = \frac{V^2}{R}$$

Where,

V is the voltage in volts,

I is the current in amperes,

R is the resistance offered by the powered devices,

T is the time in seconds

P is the power measured in watts.

Unit of Electric Power

The unit of electrical power is Watt.

If,

$$V = 1\ volts\ and\ I = 1\ ampere$$
$$P = 1\ watt$$

Thus, the power consumed in an electrical circuit is said to one watt if one ampere current flows through the circuit when a potential difference of 1 volt is applied across it. The bigger unit of electrical power is the kilowatt (kW), it is usually used in the power system

$$1\,kW = 1000W$$

Types of an Electric Power

The electrical power is mainly classified into two types. They are the DC power and the AC power:

1. DC Power

The DC power is defined as the product of the voltage and current. It is produced by the fuel cell, battery and generator.

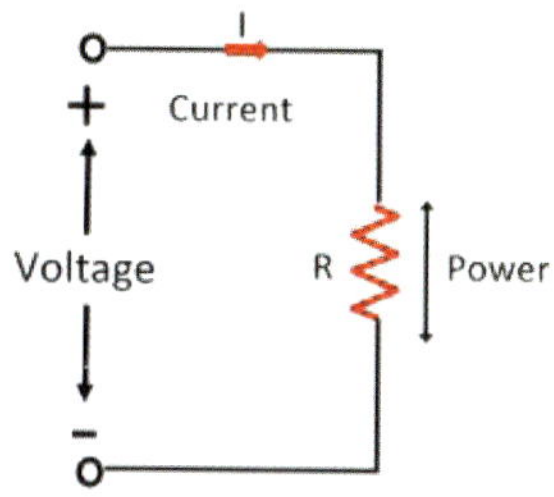

Electric power

$$P = V \times 1$$

Where,

P – Power in watt.

V – voltage in volts.

I – current in amps.

2. AC Power

The AC power is mainly classified into three types. They are the apparent power, active power and real power.

- Apparent Power – The apparent power is the useless power or idle power. It is represented by the symbol S, and their SI unit is volt-amp.

$$S = V_{rms} I_{rms}$$

Where

S – apparent power

V_{rms} – RMS voltage = $V_{peak}\sqrt{2}$ in volt.

I_{rms} – RMS current = $I_{peak}\sqrt{2}$ in the amp.

- Active Power – The active power (P) is the real power which is dissipated in the circuit resistance.

$$P = V_{max} I_{max} \, Cos\,\phi$$

Where,

P – the real power in watts.

V_{rms} – RMS voltage = $V_{peak}\sqrt{2}$ in volts.

I_{rms} – RMS current = $I_{peak}\sqrt{2}$ in the amp.

Φ – impedance phase angle between voltage and current.

- Reactive Power – The power developed in the circuit reactance is called reactive power (Q). It is measured in volt-ampere reactive.

  $$Q = V_{rms} I_{rms} \, Sin\,\phi$$

Where,

Q – the reactive power in watts.

V_{rms} – RMS voltage = $V_{peak}\sqrt{2}$ in volt.

I_{rms} – RMS current = $I_{peak}\sqrt{2}$ in the amp.

Φ – impedance phase angle between voltage and current.

The relation between the apparent, active and reactive power is shown below:

$$S^2 = Q^2 + P^2$$

The ratio of the real to the apparent power is called power factor, and their value lies between 0 and 1.

Electric Charge

Electric charge, basic property of matter carried by some elementary particles. Electric charge, which can be positive or negative, occurs in discrete natural units and is neither created nor destroyed.

Electric charges are of two general types: positive and negative.

Two objects that have an excess of one type of charge exert a force of repulsion on each other when relatively close together. Two objects that have excess opposite charges, one positively charged and the other negatively charged, attract each other when relatively near.

Many fundamental, or subatomic, particles of matter have the property of electric charge. For example, electrons have negative charge and protons have positive charge, but neutrons have zero charge. The negative charge of each electron is found by experiment to have the same magnitude, which is also equal to that of the positive charge of each proton. Charge thus exists in natural units equal to the charge of an electron or a proton, a fundamental physical constant. A direct and convincing measurement of an electron's charge, as a natural unit of electric charge, was first made (1909) in the Millikan oil-drop experiment. Atoms of matter are electrically neutral because their nuclei contain the same number of protons as there are electrons surrounding the nuclei. Electric current and charged objects involve the separation of some of the negative charge of neutral atoms. Current in metal wires consists of a drift of electrons of which one or two from each atom are more loosely bound than the rest. Some of the atoms in the surface layer of a glass rod positively charged by rubbing it with a silk cloth have lost electrons, leaving a net positive charge because of the unneutralized protons of their nuclei. A negatively charged object has an excess of electrons on its surface.

Electric charge is conserved: in any isolated system, in any chemical or nuclear reaction, the net electric charge is constant. The algebraic sum of the fundamental charges remains the same.

The unit of electric charge in the metre–kilogram–second and SI systems is the coulomb, equivalent to the net amount of electric charge that flows through a cross section of a conductor in an electric circuit during each second when the current has a value of one ampere. One coulomb consists of 6.24 × 1018 natural units of electric charge, such as individual electrons or protons. One electron itself has a negative charge of 1.6021766208 × 10−19 coulomb. In the centimetre–gram–second system there are two units of electric charge: the electrostatic unit of charge, esu, or statcoulomb; and the electromagnetic unit of charge, emu, or abcoulomb. One coulomb of electric charge equals about 3,000,000,000 esu, or one-tenth emu.

An electrochemical unit of charge, the faraday, is useful in describing electrolysis reactions, such as in metallic electroplating. One faraday equals 9.648533289×10^4 coulombs, the charge of a mole of electrons.

Inductance

Inductance is a property of an electrical circuit whereby a change in the current flowing through the circuit induces an electromotive force (EMF) that opposes the change in current. In electrical circuits, any electric current (*i*) produces a magnetic field around the current-carrying conductor, generating a total magnetic flux (Φ) acting on the circuit. This magnetic flux tends to act to oppose changes in the flux by generating a voltage (a back EMF) that counters or tends to reduce the rate of change in the current. The ratio of the magnetic flux to the current is called the self-inductance, which is usually simply referred to as the inductance of the circuit.

The property of inductance is tapped when using electrical devices known as *inductors*, some of which are shown here.

The property of inductance is particularly noticeable and useful in the form of electrical devices known as inductors. An inductor is often a coil of wire wrapped around a material called the *core*. When an electric current is passed through the coil, a magnetic field is formed around it, and this field causes the inductor to resist changes in the amount of current passing through it.

Inductors have various practical applications. For instance, when combined with capacitors, they are used as chokes in power supplies and to make tuned circuits for radio reception and broadcasting. Also, inductors are employed in transformers for power grids and as energy storage devices in some switched-mode power supplies.

Symbols and Units

The term 'inductance' was coined by Oliver Heaviside in February 1886.

It is customary to use the symbol L for inductance, possibly in honor of the physicist Heinrich Lenz.

In honor of Joseph Henry, the unit of inductance has been given the name henry (H):

1 H = 1 weber per ampere (Wb/A).

Inductance is a measure of the amount of EMF generated for a unit change in current. For example, an inductor with an inductance of 1 henry produces an EMF of 1 volt when the current through the inductor changes at the rate of 1 ampere per second.

Definitions: The quantitative definition of the (self-)inductance of a wire loop in SI units (webers per ampere) is:

$$L = \frac{N\Phi}{i}$$

where Φ denotes the magnetic flux through the area spanned by the loop, and N is the number of wire turns. The flux linkage $\lambda = N\Phi$ thus is:

$$N\Phi = Li.$$

There may, however, be contributions from other circuits. Consider for example two circuits C_1, C_2, carrying the currents i_1, i_2. The flux linkages of C_1 and C_2 are given by

$$N_1\Phi_1 = L_{11}i_1 + L_{12}i_2,$$
$$N_2\Phi_2 = L_{21}i_1 + L_{22}i_2.$$

According to the above definition, L_{11} and L_{22} are the self-inductances of C_1 and C_2, respectively. It can be shown that the other two coefficients are equal: $L_{12} = L_{21} = M$, where, M is called the mutual inductance of the pair of circuits.

The number of turns N_1 and N_2 occur somewhat asymmetrically in the definition above. But actually L_{mn} always is proportional to the product $N_m N_n$, and thus the total currents $N_m i_m$ contribute to the flux.

Self and mutual inductances also occur in the expression

$$W = \frac{1}{2} \sum_{m,n=1}^{K} L_{m,n} i_m i_n$$

for the energy of the magnetic field generated by K electrical circuits where i_n is the current in the nth circuit. This equation is an alternative definition of inductance that also applies when the currents are not confined to thin wires so that it is not immediately clear what area is encompassed by the circuit nor how the magnetic flux through the circuit is to be defined.

The definition $L = N\Phi / i$, in contrast, is more direct and more intuitive. It may be shown that the two definitions are equivalent by equating the time derivative of W and the electric power transferred to the system.

Capacitance

Capacitance is the ability of an object to hold an electrical charge. Capacitors are components in an electrical circuit that can store a charge and are considered one of the three fundamental electronic components along with inductors and resistors.

This technology doesn't measure the amount of energy stored but rather the potential energy that can be stored. This potential is measured in units called farads (F), however the capacitors in most consumer-level devices deal with thousandths of a farad (microfarads, μF) or smaller.

A capacitor consists of two conducting plates separated by an insulating layer called a dielectric. Capacitors often differ in the size and arrangements of plates as well as the type of dielectric materials used. When a capacitor is connected in a circuit, the power source's voltage forces electrons onto the surface of one plate and pulls electrons off the surface of the other plate. This results in an energy potential difference between the two plates.

The capacitance value of a capacitor can be fixed or adjusted and varying based on application. The size, shape, and proximity of the electrode plates can all affect the capacitance capability of the component. Larger plates closer together will have greater capacitance than small plates father away.

Capacitance is becoming more integrated in many applications such as computer memory, condenser microphones, radio receivers, pulsed magnets, and pressure sensors. Capacitance-based pressure transducers are extremely accurate (±0.07% full scale) when compared to sensors that use other technologies.

Resistance

The electrical resistance of an object (or material) is a measure of the degree to which the object opposes an electric current passing through it. Discovered by Georg Ohm in 1827, electrical resistance shares some conceptual parallels with the mechanical notion of friction. The SI unit of electrical resistance is the ohm, symbolized as Ω. The reciprocal quantity of electrical resistance is electrical conductance, measured in Siemens.

The resistance of an object determines the amount of current through the object for a given potential difference (voltage) across the object. Thus, electrical resistance is equal to the ratio of voltage divided by electric current. For a wide variety of materials and conditions, the electrical resistance does not depend on the amount of current passing through the object or the amount of voltage across the object. In other words, the resistance R is constant for a given material.

All materials resist current flow to some degree. They fall into one of two broad categories:

- Conductors: Materials that offer very little resistance where electrons can move easily. Examples: silver, copper, gold and aluminum.
- Insulators: Materials that present high resistance and restrict the flow of electrons. Examples: Rubber, paper, glass, wood and plastic.

Gold wire serves as an excellent conductor.

Resistance measurements are normally taken to indicate the condition of a component or a circuit:

- The higher the resistance, the lower the current flow. If abnormally high, one possible cause (among many) could be damaged conductors due to burning or corrosion. All conductors give off some degree of heat, so overheating is an issue often associated with resistance.
- The lower the resistance, the higher the current flow. Possible causes: insulators damaged by moisture or overheating.

Many components, such as heating elements and resistors, have a fixed-resistance value. These values are often printed on the components' nameplates or in manuals for reference.

When a tolerance is indicated, the measured resistance value should be within the specified resistance range. Any significant change in a fixed-resistance value usually indicates a problem.

"Resistance" may sound negative, but in electricity it can be used beneficially.

References

- Electronic-devices-and-circuits: physics-and-radio-electronics.com, Retrieved 10 May 2018
- What-is-voltage: theengineeringmindset.com, Retrieved 03 March 2018
- Electric-charge, science: britannica.com, Retrieved 23 April 2018
- What-is-capacitance: setra.com, Retrieved 31 March 2018
- Electrical-resistance: newworldencyclopedia.org, Retrieved 30 June 2018

Branches of Electronics

The field of electronics has a number of significant branches, such as digital and analogue electronics, microelectronics, power electronics, optoelectronics, etc. This chapter closely examines the fundamental principles and concepts related to these fields for a holistic understanding of electronics.

Digital Electronics

Digital Electronics is the sub-branch of electronics which deals with digital signals for processing and controlling various systems and sub-systems. In various applications like sensors and actuators, usage of digital electronics is increasing extensively.

Digital Signals

Digital electronics is entirely the field in which digital signals is used. Digital signals are discretization of analog signals. A signal carries information. In digital signals the values in a particular band is same i.e constant. Digital signals form the basis of digital circuit and digital electronics.

You must be thinking that why we need digital signals when we have analog signals and analog circuits. The reason behind this is analog signals are more likely to get affected by noise, while digital signals are noise immune. Besides, the designing and operations of a digital circuit is simpler than analog circuit.

Thus, it is better to use digital signals if we need accurate and precised operation of the devices. Now, the question emerges that why digital signals are noise immune. What makes them perform better in noise interference. To understand this let's consider both the signals analog signals and digital signals.

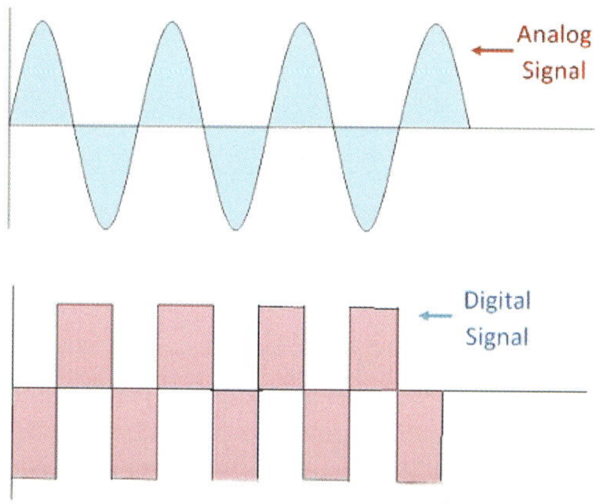

It is evident from the above diagram that analog signal is a continuous tome varying signal while digital signal consists of discrete values of voltage and thus it do not varies continuously with time.

Design

Field of Digital Electronics is Binary Arithmetic and Boolean algebra.

Basic Logic Gates

Logic gates constitute the foundation blocks for digital logic. Let us start by reviewing these gates and their truth tables.

An AND Gate has two or more inputs and produces one output as follows: output = 1 if all of the inputs are high, output = 0 if one or more of the inputs are low:

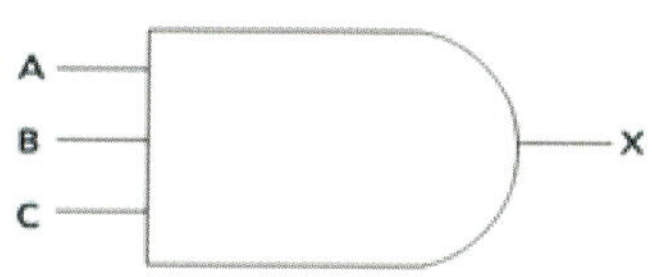

Inputs			Output
C	B	A	X
0	0	0	0
0	0	1	0
0	1	0	0
0	1	1	0
1	0	0	0
1	0	1	0
1	1	0	0
1	1	1	1

An OR gate also has two or more inputs and produces one output as follows: output = 1 if one or more inputs are high, output = 0 if all inputs are low.

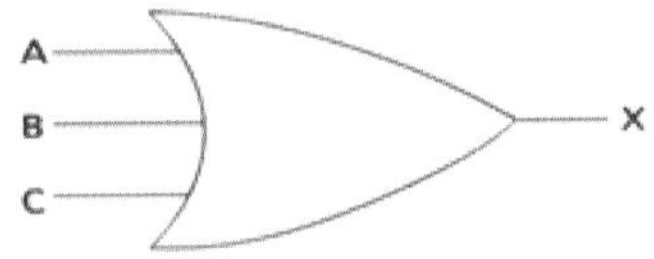

Inputs			Output
C	B	A	X
0	0	0	0
0	0	1	1
0	1	0	1
0	1	1	1
1	0	0	1
1	0	1	1
1	1	0	1
1	1	1	1

The inverter gate has one input and produces one output as follows: output =1 if input is low, output = 0 if input is high.

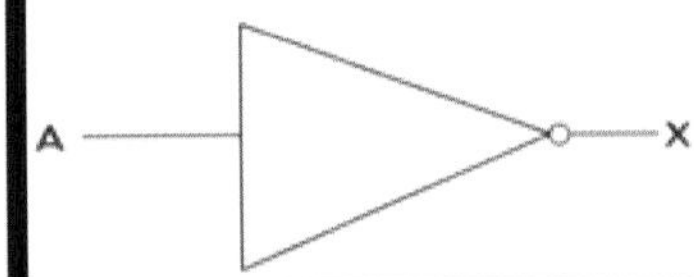

Inputs	Output
A	X
1	0
0	1

The NAND gate has two or more inputs and produces one output as follows: output = 0 if all the inputs are high, output = 1 if any of the inputs are low.

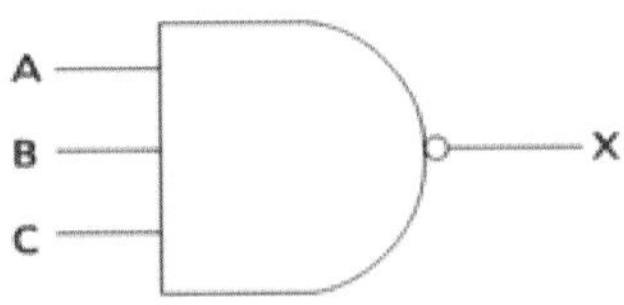

Inputs			Output
C	B	A	X
0	0	0	1
0	0	1	1
0	1	0	1
0	1	1	1
1	0	0	1
1	0	1	1
1	1	0	1
1	1	1	0

The NOR gate has two or more inputs and produces one output as follows: output = 1 if all inputs are low, output = 0 if any of the inputs is high.

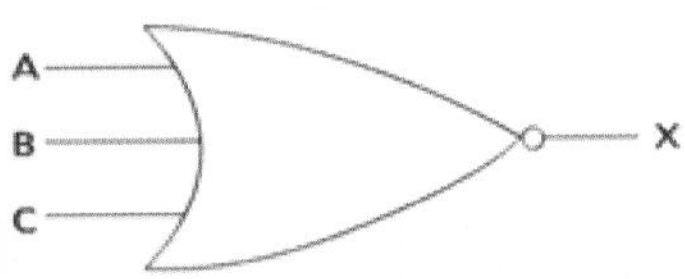

Inputs			Output
C	B	A	X
0	0	0	1
0	0	1	0
0	1	0	0
0	1	1	0
1	0	0	0
1	0	1	0
1	1	0	0
1	1	1	0

The Exclusive-OR gate always has two inputs only and produces one output as follows: output = 1 when inputs are not similar, output = 0 when inputs are the same.

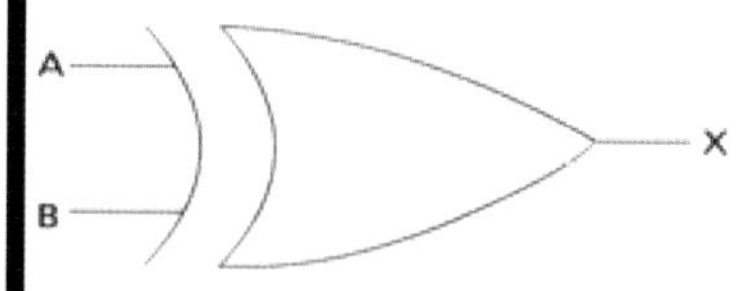

Inputs		Output
B	A	X
0	0	0
0	1	1
1	0	1
1	1	0

The Exclusive-NOR gate always has two inputs only and produces one output as follows: output = 1 when inputs are both high or are both low, output = 0 when inputs are not similar.

Inputs		Output
B	A	X
0	0	1
0	1	0
1	0	0
1	1	1

Boolean Algebra

Boolean algebra has similar rules to other algebras and these rules are used to manipulate the expression at hand. Some of these basic rules are.

One variable:

A. A = A

A + A = A

One variable and 0 or 1:

A. 0 = 0

A. 1 = A

A + 0 = A

A + 1 = 1

DeMorgan's Theorem:

(A + B)' = A'. B'

(A. B)' = A' + B'

Associative:

(A . B) . C = A. (B. C)

(A +B) + C = A + (B + C)

Commutative:

A. B = B . A

A + B = B + A

Distributive:

A. (B + C) = A. B + A. C

A + (B. C) = (A + B). (A + C)

Note: OR operator is represented by the '+' symbol and has the lowest precedence. NOT operator is represented by the " ' " symbol and has the highest precedence. AND operator is represented by '.' symbol and is in the middle.

Example Exercise

STEP 1: Simplify the following expression: ABC'D + AB'C'D + A'BCD + AB'CD + A'BCD'.

Answer Sub-Step 1: Using the above rules we can simplify the expression as follows:

ABC'D + AB'C'D + A'BCD + AB'CD + A'BCD'

= ABC'D + AB'C'D + A'BCD + A'BCD'+ AB'CD

= AC'D(B+B') + A'BC(D+D') + AB'CD

= AC'D(1) + A'BC(1) + AB'CD

= AC'D + A'BC + AB'CD

= AC'D + AB'CD + A'BC

= AD(C'+B'C) + A'BC

= AD((C'+B')(C'+C)) + A'BC

= AD((C'+B')(1)) + A'BC

= AD(BC)' + A'BC

Example Problem

STEP 2: Using NI Multisim, open circuit file "logic_gates.ms12". You will notice the circuit below:

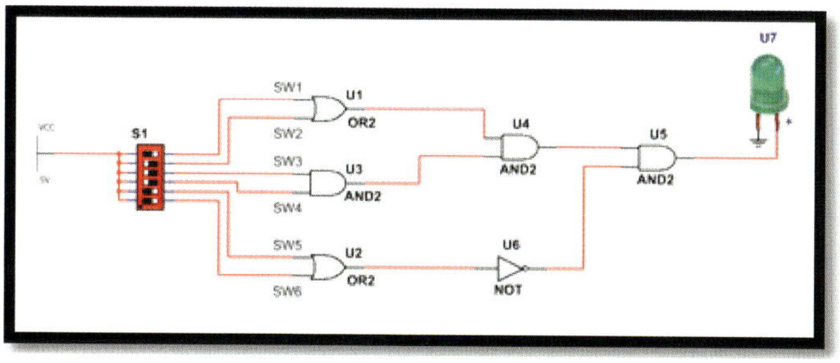

The above circuit shows the different connections of logic gates used to control a machine (indicated by an LED) in a workshop. There are On/Off switches, micro-switches, as well as emergency stop switches available (SW1 &SW2, SW3 & SW4, SW5 & SW6 respectively).

Answer Sub-step 2: Based on the circuit above, determine the switch combination that will turn the machine on.

Based on the individual truth tables for each of the logic gates, we start from the far right working our way back to determine the combination of bits or switches that will enable the LED to turn on or in other words the output of 'U5' to be equal to 1:

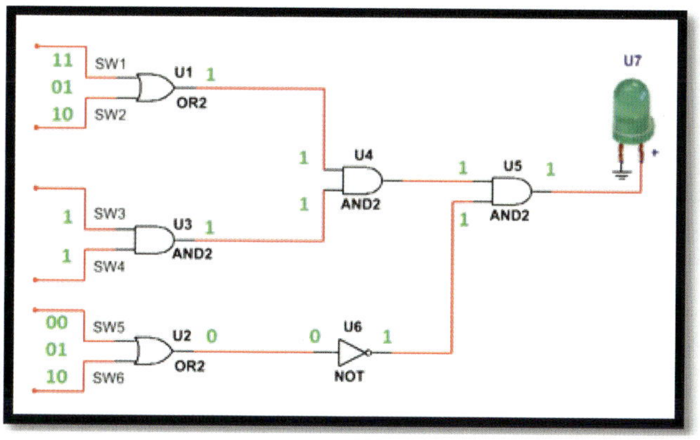

Input			Output
SW1 & SW2	SW3 & SW4	SW5 & SW6	LED
11	11	00	1
11	11	01	1
11	11	10	1
01	11	00	1
01	11	01	1
01	11	10	1
10	11	00	1
10	11	01	1
10	11	10	1

STEP 3: Using circuit file "logic_gates.ms12", toggle the switches to replicate our findings and verify our combinations. You will notice that the LED will turn on when the above combinations are selected as shown below:

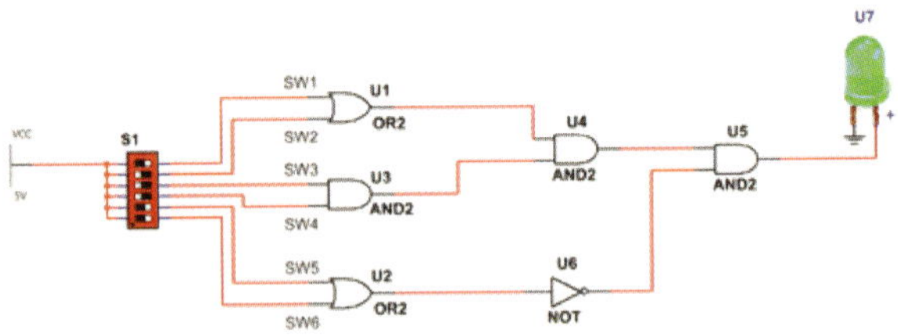

Advantages of Digital Systems

- Easier Designing: The Digital systems can be easily designed as they involve digital signals. These signals do not require exact value at a particular time but it consists of range of particular values of voltage. Thus, it comprises of basically two values 0 and 1 i.e high or low.
- Noise Immune: Digital systems are noise immune because digital signal consists of range of particular values. Thus, when noise is introduced in the medium and digital signal and analog signal both passes through it. The analog signal will be affected more because it varies continuous with time so it is difficult to identify that noise has destroyed which value of voltage. While in case of digital system, noise effect the particular range of the signal thus, it is clear to identify the particular range of filtering is also easy in case of digital signals.
- Information Storage is Simpler: The storage of information in digital systems is easy. It can be stored by latching thus, the it can be stored for a long period of time.
- High Accuracy And Precision: The digital signal offers high accuracy and precision. This is because the processing of digital signal is done through the switching circuit. While in case of analog signals the processing and its output is highly dependant on circuit components. The accuracy obtained in analog circuits is restricted to 3-4 digits while in case of digital signals the accuracy is far more than the analog circuit.
- Programmable: The digital systems are easily programmable but analog system becomes complex when excessive programming of components is done.

Disadvantages of Digital Systems

Everything comes up with pros and cons. Similarly digital systems too hold certain disadvantages which are as follows:

- Expensive: Digital systems are expensive because it involves switching elements.

- Analog nature of Real World Entities: We need to convert the digital output in analog form because all the real world entities are analog.

Digital to Analog and Analog to Digital Conversion

These types of conversion circuits are required so that we can convert analog signal into digital and then process the digital signal using digital systems and convert back the digital signal into analog form so that it is can be understood by real world.

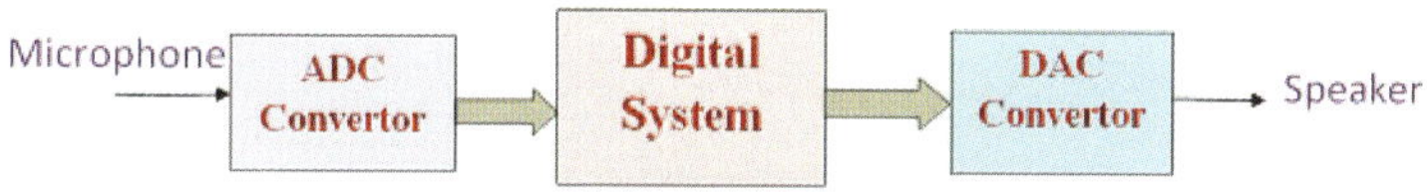

Conversion System

In this way we can convert one form of signal into another form and process it digitally.

Analog Electronics

Analog electronics is a branch of electronics that deals with a continuously variable signal. It's widely used in radio and audio equipment along with other applications where signals are derived from analog sensors before being converted into digital signals for subsequent storage and processing. Although digital circuits are considered as a dominant part of today's technological world, some of the most fundamental components in a digital system are actually analog in nature.

In order to understand the concept, let's first try to analyze the word 'Analog'.

Analog

Analog means continuous and real. The world we live in is analog in nature, implying that it's full of infinite possibilities. The number of smells we can sense, the number of tones we can hear, or the number of colors we can paint with; everything is infinite. The people working in the field of analog electronics are basically dealing with analog devices and circuits.

Analog

For example, if we build a circuit and it counts values like 1, 2, 3, 4, and 5; the values are neither infinite nor continuous. On the other hand, if the circuit counts like 1.00000, 1.00001, 1.00002, 4.99999, 5.00000, the amount of information would be infinite.

Analog Signals

Before proceeding with analog signals, let's understand the simple meaning of a signal. In electrical engineering, signals are basically time-varying quantities (usually voltage or current). So when we talk about the signal it means we are talking about a voltage that's changing over time.

Signals are passed among devices in order to obtain or send information in the form of audio, video or encoded data. The transmission takes place through wires or via air through radio frequency waves. For instance, audio signals are transferred from the computer's audio card to the speakers, while data signals between a tablet and a Wi-Fi router pass through the air.

Analog Signals use attributes of the medium to convey the signal's information. For example, an aneroid barometer makes use of the angular position of a needle to convey the changes in atmospheric pressure. The signals take any value from a given range and each signal value denotes different information. Each level of the signal signifies a different level of the phenomenon and any change in the signal is meaningful.

It's quite easy to determine if a signal graph is analog or digital. The former one is smooth and continuous while the latter is edge and appears in the form of stepping squares. Given below is an analog signal graph representing the change in voltage with the change in time.

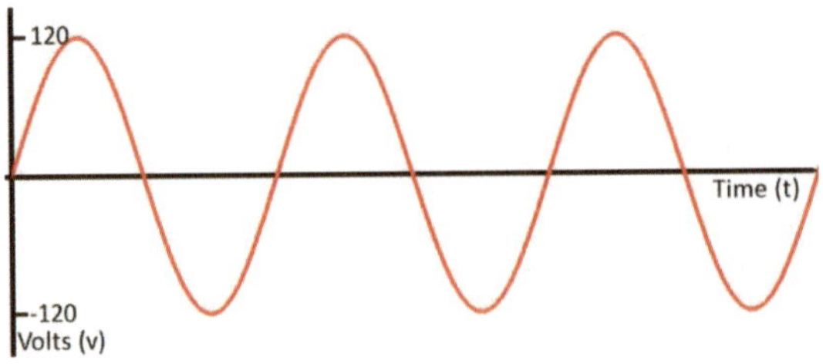

An analog signal graph representing time-voltage graph

Analog Circuits

Analog circuits can be defined as a complex combination of op amps, resistors, caps, and other basic electronic components. These circuits can be as simple as a combination of two resistors to make a voltage divider or elegantly built with many components. Such circuits can attenuate, amplify, isolate, modify, distort the signal, or even convert the original one into a digital signal.

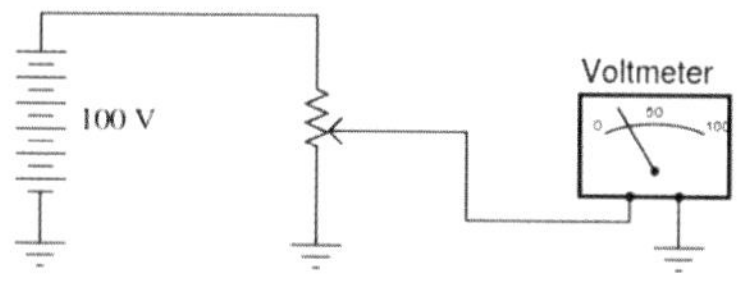

Analog Circuit

These circuits are difficult to design and need a lot of precision as compared to the digital circuits. The modern circuits are rarely found to be completely analog as these days analog circuitry may use digital or microprocessor techniques to improve performance. Such circuits are called mixed signals.

There are two types of analog circuits namely passive and active, where the former ones don't consume any electrical power while the latter ones do.

Using Analog Components and Circuitry

As mentioned above, analog components are the most basic type of components while the digital components are widely used in today's devices. Both the branches are independently significant in their own way so we are enlisting certain cases when the analog components and circuitry would be ideal.

Filtering Signals: While dealing with a continuous signal, a continuous analog filter is needed to remove all possible frequency content that you don't want. As compared to a digital filter, it is much easier to use as well as costs less.

High Power: Although digital measurement and control may be useful for high power systems, a digital signal switching from 0 to 400V may not be efficient enough. The continuous nature of power delivery in the AC and DC systems require analog components as these are more durable and characterized.

Before A/D and After D/A: In order to switch between continuous and discrete data, hybrid systems need both Analog-to-Digital and Digital-to-Analog converters, but as per the Nyquist's Theorem, the sampling frequency must be at least twice the frequency of the highest frequency component contained in the signal. So, in order to fulfill the theorem, any signal inadvertently included in the original signal can be filtered so as to remove noise after sampling.

Since the signal is not yet digital, the only option is to filter the signal with an analog unit. Once you are done operating on a signal digitally and convert it back to analog, all the processing must once again be done with analog components and circuitry.

Sensors: Sensors convert the real world information into data that can be recognized by a computer or an embedded system. Often the information is not readily available in the real world; rather the sensors first create an analog signal and later convert it into digital signals. Unlike the high voltage systems, sensors have low amplitude and need signal conditioning to increase the value of the signal so as to make a better use of the full range of an ADC.

Analog vs Digital Electronics

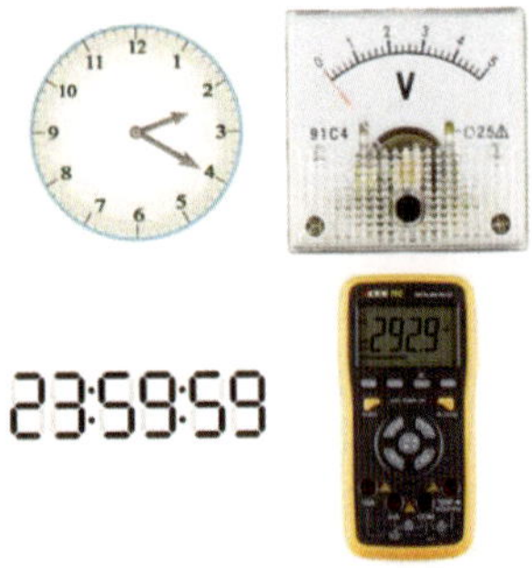

Analog vs Digital Electronics

The most basic difference between analog and digital electronics is that in the former technology translates the information into electric pulses of varying amplitude, while the latter translates information into a binary format of 0 and 1, where each bit represents two distinct amplitudes.

However, there are certain characteristic features that clearly state the difference between analog and digital electronics. Let's have a look at these key aspects:

Factors	**Analog**	**Digital**
waves	Denoted by Sine waves	Denoted by Square waves
Signal	Continuous signal representing physical measurements	Discrete signal representing discrete time signals generated by digital modulation
Data Transmission	Subject to deterioration by noise	Noise-immune without deterioration
Bandwidth	Consumes less bandwidth	Consumes more bandwidth
Memory	Stored in the form of wave signal	Stored in the form of binary bit
Power	Draws large power	Draws negligible power
Impedance	Low impedance	High order of 100 megaohm
Errors	Analog instruments have considerable observational errors	Digital instruments are free from observational errors

Microelectronics

Microelectronic is branch of electronic technology devoted to the design and development of extremely small electronic devices that consume very little electric power.

It originated in the early 1960's as a result of the increasing size and complexity of functions of electronicapparatus and the more stringent reliability requirements. The use of several thousands, and even tens of thousands, ofindividually fabricated electron tubes, transistors, capacitors, resistors, transformers, and other elements in certain apparatusand connection of the leads by soldering or welding rendered apparatus unwieldy, labor-intensive in assembly, andinsufficiently reliable; it also caused high power consumption. The search for methods of eliminating these disadvantages ledto the appearance of new trends in the design of electronic apparatus, such as printed circuits, modules, micromodules, andintegrated circuits (based on multiple processing methods).

Through the use of advances in solid state physics, and particularly semiconductor physics, microelectronics is solving theproblems mentioned above not merely by reducing the size of the electronic elements but by creating electronic structures (functional units and assemblies) that are structurally, technologically, and electrically integrated. In such structures a largenumber of microminiature elements and their electrical connections are structurally integrated according to the circuit in acommon technological process. This process, which was made possible by the planar process of making semiconductordevices introduced in 1959, involves the use of an initial common blank (usually in the form of wafers of a semiconductormaterial) for a large number (100–2,000) of identical functional electronic assemblies, which pass simultaneously through aseries of technological operations under identical conditions. Thus, each such assembly is produced as a result of step-by-step integrated multiple processing of many identical assemblies on the same wafer, rather than through the assembly ofdiscrete elements. During the manufacturing process, the properties of various circuit elements and their combinations areimparted to certain parts of the semiconductor material, thus producing the overall assembly. This microminiature assembly, after being detached from the wafer and mounted in a housing, is called an integrated microcircuit, or integrated circuit. Inthis connection the concept of a circuit element itself is altered in microelectronics. In practical terms, the integrated circuit asan indivisible unit consisting of five or more elements becomes an element. An integrated circuit is characterized by the levelof integration—that is, by the number of simpler elements in it.

In view of the exceptionally high accuracy in carrying out the technological processes and the large number of operationsinvolved in the fabrication of microelectronic products, a variety of high-quality semiconductor and other materials, as well asprecision equipment is required. Monocrystalline silicon serves as the semiconductor base material. The equipment mustprovide for production of the elements within a dimensional tolerance of microns and fractions of a micron.

Several overlapping and complementary trends may be distinguished in microelectronics, depending on the design, technology and physical principles used. Among the trends are integrated electronics, vacuum microelectronics, opticalelectronics, and functional electronics. Integrated electronics has achieved the highest level of development. Its appearanceopened extensive possibilities for microminiaturization of radioelectronic devices and began the process of creating third-generation apparatus, using integrated circuits (first-generation apparatus used vacuum tubes; second-generation apparatus, semiconductor devices). The area of use of integrated circuits extends from computers and systems for space flight tohousehold appliances. The rate of growth of integrated-circuit production is very high. More than 1 billion integrated circuits were manufactured worldwide in 1972.

Semiconductor integrated circuits were first produced on the basis of methods of multiple processing, by forming the requirednumber of electronic elements and electrical connections inside one semiconductor crystal. They are made mainlyby means of planar-epitaxial technology, which was borrowed from the production of discrete semiconductor devices anddiffered only in the additional operations required for the electrical insulation of the individual elements on the semiconductorwafer and the connection of all elements in the crystal to form a functional assembly. Insulation is produced by creatingaround the element an area of semiconductor material with the opposite type of conductivity (thereby forming an insulating p-n junction) or a layer of a dielectric, such as silicon dioxide. The principal operations of planar-epitaxial technology are themechanical and chemical processing of a semiconductor wafer, the epitaxial buildup on the wafer of a layer with the requiredelectrophysical properties (type of conductivity, resistivity, and so on), photolithography, doping (by diffusion or ionimplantation), and the deposition of metallic films for electrodes, connecting paths, and contact areas.

Photolithography is the most critical of the processing steps listed above. It provides for the selective processing of certainportions of the semiconductor wafer, such as the etching of a "window" in the oxide film on the wafer to permit the diffusion ofimpurities. A light-sensitive lacquer called a photoresist is used in this process. A film of the photoresist is deposited on thewafer and then is exposed to ultraviolet light through a photographic mask ("phototemplate"), which is a glass plate with arecurrent pattern formed by opaque and semitransparent areas (usually a layer of chromium) and pressed firmly to the wafer.After irradiation, the photoresist film undergoes selective etching, which causes reproduction of the patterns of thephototemplate on the wafer. The photoresist can also be irradiated by a contactless method, in which the pattern is projectedonto the wafer. A promising method of exposing the wafer to the pattern is the electron-beam process (electronic photolithography).

In the production of semiconductor integrated circuits the photolithographic process must be repeated to reproduce varioussuperimposed patterns on the wafer. A set of seven or eight phototemplates is usually used. The design and production ofthe phototemplates require particularly high accuracy and the maintenance of airtight cleanliness conditions in the productionshops (no more than three to five dust particles about 0.5 micron in size per liter of air): to produce hundreds of elementshaving dimensions measured in microns for hundreds of identical integrated circuits that are fabricated simultaneously on asingle wafer, the phototemplates must ensure the reproducibility of the dimensions from one pattern to another and theirmutual coincidence.

Consequently, complicated precision equipment, such as coordinatographs with computer programcontrol to draw the original pattern magnified hundreds of times and photographic dies of various designs to reduce andduplicate the original pattern, is used in the design and production of phototemplates.

To form the structures of the elements in the original semiconductor wafer, the areas prepared in the photolithographic stageare doped with impurities. The main method of doping is diffusion, in which the silicon wafer is exposed for a certain time tovapors of the impurity at a temperature of 1100°-1200°C. The accuracy with which the temperature is maintained the steadiness of the impurity concentration near the surface, and the duration of the process control the distribution of theimpurity with respect to the thickness of the wafer and, correspondingly, the parameters of the element being formed. In addition to diffusion, doping may be done by ion implantation (bombarding the wafer with ionized impurity atoms), a newtechnological trend that supplements and partially replaces diffusion. Semiconductor integrated circuits have a high level of integration (up to 10,000 elements, and sometimes more, in one crystal).

The advancement of the technology for producing active elements (diodes and transistors) on wafers of semiconductormaterial through the transition to multiple processing has stimulated the development of printed circuitry and film technologyto produce passive microminiature components (resistors and capacitors) that formed the basis for the development of filmintegrated circuits. Such circuits are usually purely passive, because the deposition of monocrystalline semiconductor films toform active elements does not provide the required quality. The base for a film integrated circuit is a dielectric substrate (forexample, a ceramic). The technology of production of such circuits is divided into the thick-film type, in which layers ofconducting, resistive, and dielecric pastes are applied in thicknesses of 1–25 microns (μ), and the thin-film type, in which films up to 1 JUL thick are sputtered in a vacuum through metal stencils, or sputtering is used in combination withsubsequent photolithographic processing.

A film integrated circuit with unencapsulated individual semiconductor devices (diodes and transistors) and unencapsulatedsemiconductor integrated circuits mounted on it is known as a hybrid integrated circuit. Its passive portion may bemultilayered, in the form of a set of ceramic substrates with layers of film circuit elements. Sintering of the substratesproduces a monolith with a multilayered arrangement of electrically interconnected passive elements. The unencapsulatedactive elements are mounted on the upper surface of the monolith.

In addition to semiconductor and film integrated circuits, a combined type is being produced. The active elements aremanufactured in the body of the semiconductor substrate by planar-epitaxial technology, and the passive elements andelectrical connections are applied to the surface of the monolithic structure in the form of thin films. The level of integration ofcombined integrated circuits approaches that of semiconductor circuits.

Multichip integrated circuits are also produced with a high level of integration, in which several semiconductor integrated-circuit crystals are combined on a dielectric substrate with film connections in a very complicated electronic device. Itsfunctional purpose may correspond to that of an individual unit or even a system, such as a desk calculator.

The combination of the film technology of producing passive elements with the use of microminiature electrovacuum devicesas the active elements has resulted in the appearance of vacuum integrated circuits and a new field, vacuummicroelectronics. A vacuum integrated circuit may be made in the form of a film integrated circuit with microminiatureelectrovacuum devices mounted on it or in the form of apparatus having all its components in a vacuum. Unlikesemiconductor integrated circuits, vacuum circuits have greater resistance to the effects of cosmic radiation; their packingdensity is up to 20–30 elements per cm^3.

All types of integrated circuits are divided into two large groups, depending on their functional attributes—digital (logic) and linear circuits. Digital circuits are designed to operate in logic units, particularly in digital computers. All other integrated circuits belong to the linear class and are designed mainly for the linear processing (amplification, modulation, detection, and so on) of electrical signals, although they may also include such nonlinear elements as sine-wave generators and frequency converters.

Microelectronics is developing in two main directions: an increase in the level of integration and the packing density inintegrated circuits, which has become traditional, and the search for new physical principles and phenomena for the development of electronic devices that perform the functions of a circuit or even a system. The first trend has resulted inlevels of integration characterized by many thousands of elements in one circuit housing, with micron and submicron dimensions of certain elements. The second trend may make it possible to abandon further increases in the level ofintegration (because of structural complexity), as well as to reduce the dissipation of power and increase the speed-of-response of the apparatus. In general, this new trend is called functional microelectronics, or the electronics of combined media, using such phenomena as solid-state optical phenomena (optical electronics), the interaction of an electron flux with acoustic waves in solids (acoustic electronics), the properties of superconductors, and the properties of magnetic materials and semiconductors in magnetic semiconductors (magnetoelectronics).

Power Electronics

Power electronics is the technology that acts as an interface between the electrical source and the electrical load. The electrical source is the power supply available and the load is the appliance that needs to be powered. This technology ensures that power is taken from the source to the load in the most efficient and robust way.

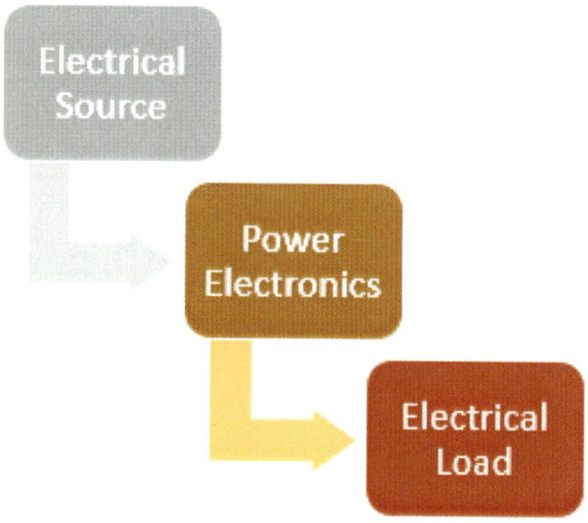

The power electronics interface essentially consists of a controller and a converter. The converter transforms the energy into desired form whereas the controller sets the magnitude of required voltage or current level for end-use with respect to a fixed reference value.

The power electronics components that are typically used are diodes, MOSFETs (high-frequency switches), inductors, capacitors and the like.

A typical example of power electronic equipment is the battery charger for mobile phones. Here the input AC voltage is first converted into DC voltage and then regulated to a lower DC voltage in order to charge the battery. All the conversion process is done by the battery charger which comprises of power electronic circuits. This is just one basic example of how much a role power electronics plays in our daily lives. Variable-speed drives for motors and even regulators for domestic fans are examples of power electronics applications.

Power Electronics involves the study of:

- Power semiconductor devices - their physics, characteristics, drive requirements and their protection for optimum utilisation of their capacities,
- Power converter topologies involving them,
- Control strategies of the converters,
- Digital, analogue and microelectronics involved,
- Capacitive and magnetic energy storage elements,
- Rotating and static electrical devices,
- Quality of waveforms generated,
- Electro Magnetic and Radio Frequency Interference,
- Thermal Management.

Devices

Power electronics devices which act as solid-state switches in the circuits. They act as a switch without any mechanical movement:

- Power Diodes
- Metal-Oxide-Semiconductor Field-Effect Transistor (MOSFET)
- Bipolar -Junction Transistor (BJT)
- Insulated-Gate Bipolar Transistor (IGBT)
- Thyristors (SCR, GTO, MCT)

Solid-state devices are completely made from a solid material and their flow of charges is confined within this solid material. This name "solid state" is often used to show a difference with the earlier technologies of vacuum and gas-discharge tube devices; and also to exclude the conventional electro-mechanical devices (relays, switches, hard drives and other devices with moving parts).

The transistor by Bell Labs in 1947 was the first solid-state device to come into commercial use later in the 1960s. In this article, similar solid-state devices such as power diode, power transistor, MOSFET, thyristor and its two-transistor model, triac, gate turn-off thyristor (GTO), insulated-gate bipolar transistor (IGBT) and their characteristics (such as i-v characteristics and turn-off characteristics) is also presented. In power electronics circuitry, these switches act in saturation region and work in linear region in the analog circuitry such as in power amplifiers and linear regulators. This makes these switches highly efficient since there are lesser losses during the power processing.

Power Diode

A power diode has a P-I-N structure as compared to the signal diode having a P-N structure. Here, I (in P-I-N) stands for intrinsic semiconductor layer to bear the high-level reverse voltage as compared to the signal diode. However, the drawback of this intrinsic layer is that it adds noticeable resistance during forward-biased condition. Thus, power diode requires a proper cooling arrangement for handling large power dissipation. Power diodes are used in numerous applications including rectifier, voltage clamper, voltage multiplier and etc. Power diode symbol is the same as of the signal diode as shown in figure.

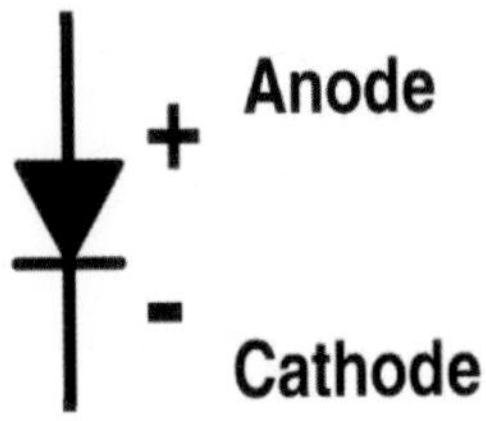

Figure: Symbol for Power Diode

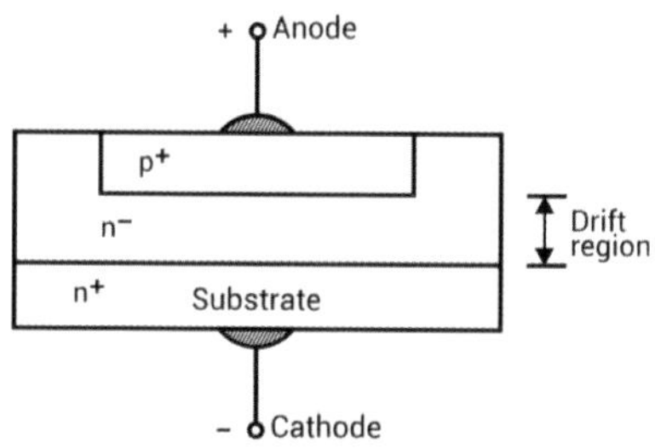

Figure: Structure of Power Diode

Other features that are incorporated in the power diode letting it handle larger power are:

(a) Use of guard rings.

(b) Coating of Silicon Dioxide layer.

Guard rings are of p-type that prevents their depletion layer merge with the depletion layer of the reverse-biased p-n junction. The guard rings prevent the radius of the curvature of the depletion layer boundary to become too narrow which increases the breakdown strength. Coating of SiO_2 layer helps in limiting the electric field at the surface of the power diode.

If thickness of lightly doped I layer (n- layer) > depletion layer width at breakdown ⇒ Non-punch through power diode.

(This means depletion layer has not punched through the lightly-doped n-layer).

If thickness of I layer < depletion layer width at breakdown ⇒ Punch through power diode.

Characteristics of Power Diode

The two types of characteristics of a power diode are shown in figure below are named as follows:

1. Amp-volt characteristics *(i-v characteristics)*
2. Turn-off characteristics *(or reverse-recovery characteristics)*

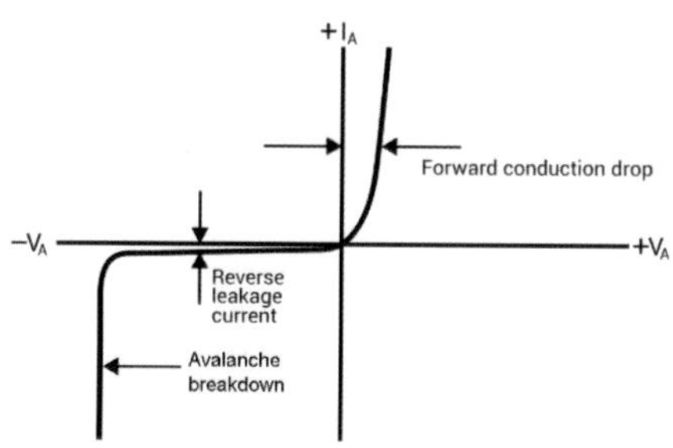

Figure: Amp-Volt Characteristics of Power Diode

Cut-in voltage is the value of the minimum voltage for VA (anode voltage) to make the diode works in forward conducting mode. Cut-in voltage of signal diode is 0.7 V while in power diode it is 1 V. So, its typical forward conduction drop is larger. Under forward-bias condition, signal diode current increases exponentially and then increases linearly. In the case of the power diode, it almost increases linearly with the applied voltage as all the layers of P-I-N remain saturated with minority carriers under forward bias. Thus, a high value of current produces results in voltage drop which mask the exponential part of the curve. In reverse-bias condition, small leakage current flows due to minority carriers until the avalanche breakdown appears as shown in figure below.

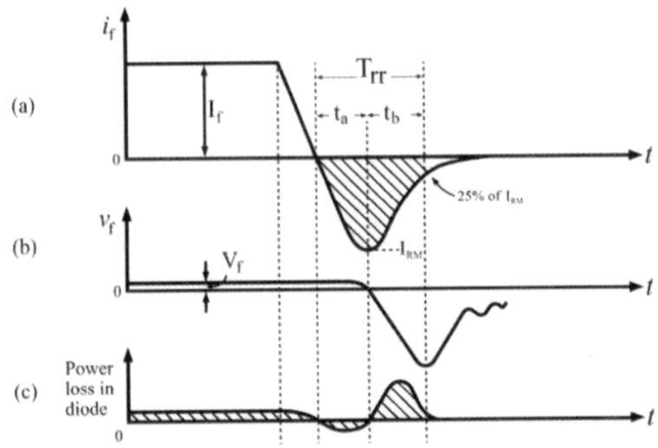

Figure: Turn-Off Characteristics of Power Diode: a) Variation of Forward Current i_f ; b) Variation of Forward Voltage Drop v_f ; c) Variation of Power Loss

After the forward diode comes to null, the diode continues to conduct in the opposite direction because of the presence of stored charges in the depletion layer and the p or n-layer. The diode current flows for a reverse-recovery time t_{rr}. It is the time between the instant forward diode current becomes zero and the instant reverse-recovery current decays to 25 % of its reverse maximum value.

Time T_a : Charges stored in the depletion layer removed.

Time T_b : Charges from the semiconductor layer is removed.

Shaded area in figure.a represents stored charges Q_R which must be removed during reverse-recovery time t_{rr}.

Power loss across diode = $v_f * i_f$.

As shown, major power loss in the diode occurs during the period t_b.

Recovery can be abrupt or smooth as shown in figure. To know it quantitatively, we can use the S – factor.

Ratio T_b/T_a : Softness factor or S-factor.

S-factor: measure of the voltage transient that occurs during the time the diode recovers.

S-factor = 1 ⇒ low oscillatory reverse-recovery process. (Soft –recovery diode)

S-factor <1 ⇒ large oscillatory over voltage (snappy-recovery diode or fast-recovery diode).

Power diodes now exist with forward current rating of 1A to several thousand amperes with reverse-recovery voltage ratings of 50V to 5000V or more.

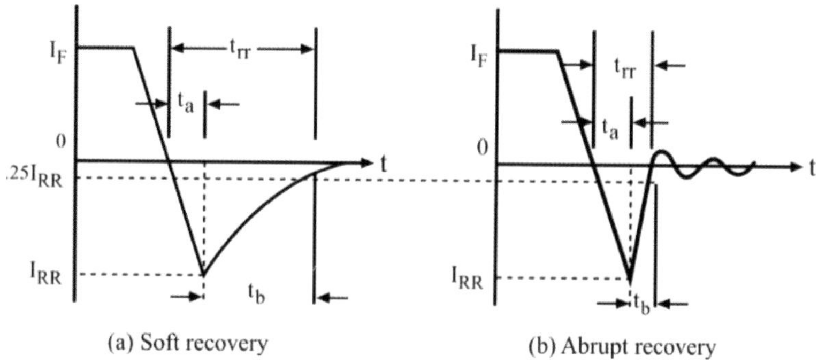

Figure: Reverse-Recovery Characteristics for Power Diode

Schottky Diode: It has an aluminum-silicon junction where the silicon is an n-type. As the metal has no holes, there is no stored charge and no reverse-recovery time. Therefore, there is only the movement of the majority carriers (electrons) and the turn-off delay caused by recombination process is avoided. It can also switch off much faster than a p-n junction diode. As compared to the p-n junction diode it has:

(a) Lower cut-in voltage;

(b) Higher reverse leakage current;

(c) Higher operating frequency.

Application: high-frequency instrumentation and switching power supplies.

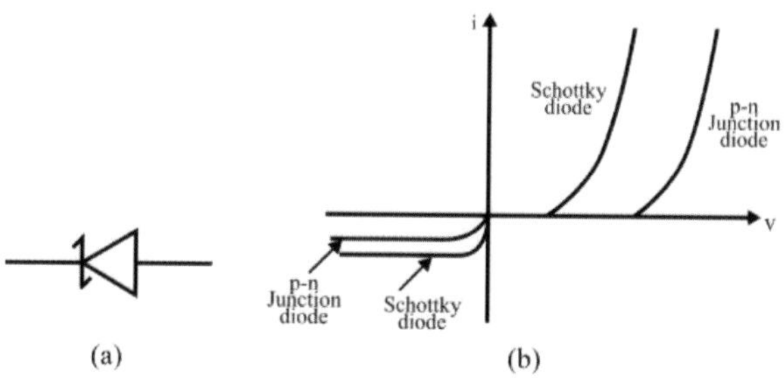

Figure: Schottky Diode Symbol and Current-Voltage Characteristics Curve

Metal-Oxide Semiconductor Field-Effect Transistor (Power)

MOSFET is a voltage-controlled majority carrier (or unipolar) three-terminal device. Its symbols are shown in figures. As compared to the simple lateral channel MOSFET for low-power signals, power MOSFET has different structure. It has a vertical channel structure where the source and the drain are on the opposite side of the silicon wafer as shown in figure. This opposite placement of the source and the drain increases the capability of the power MOSFET to handle larger power.

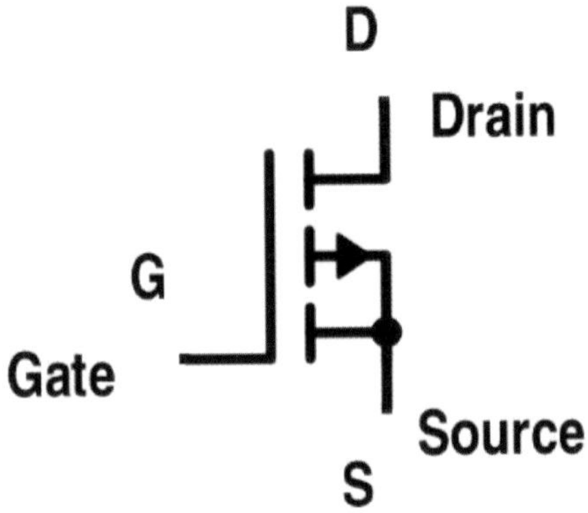

Figure: MOSFET Symbol

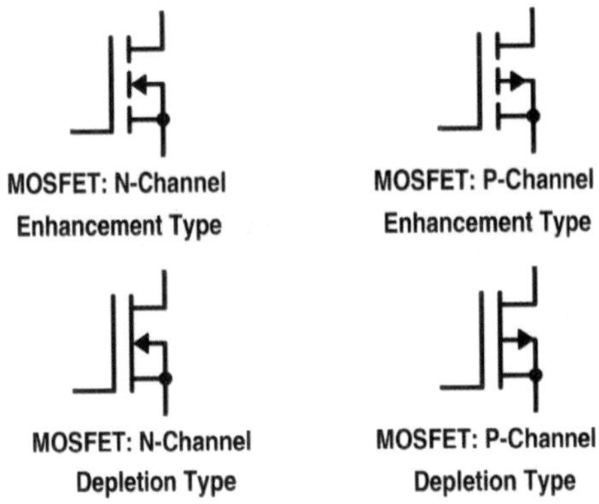

Figure: MOSFET Symbols for Different Modes

In all of these connections, substrates are internally connected. But in cases where it is connected externally, the symbol will change as shown in the n-channel enhancement type MOSFET in figure. N-channel enhancement type MOSFET is more common due to high mobility of electrons.

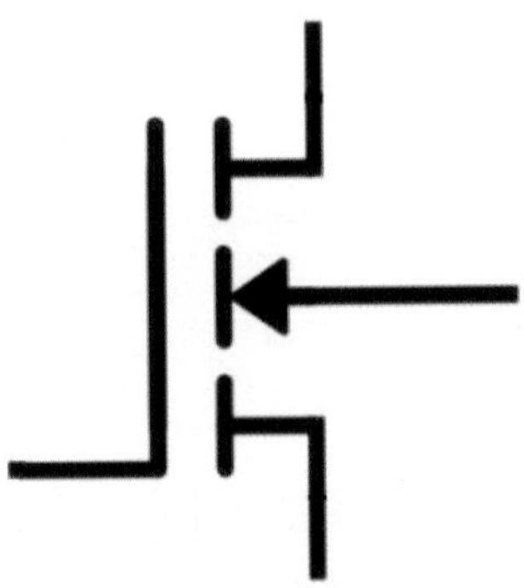

Figure: N-channel Enhancement-Type MOSFET with Substrate Connected Externally

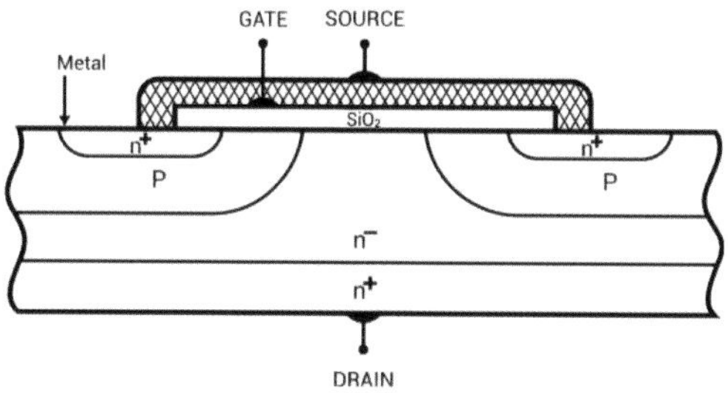

Figure: Cross-Sectional View of the Power MOSFET

Basic circuit diagram and output characteristics of an n-channel enhancement power MOSFET with load connected are in figures.

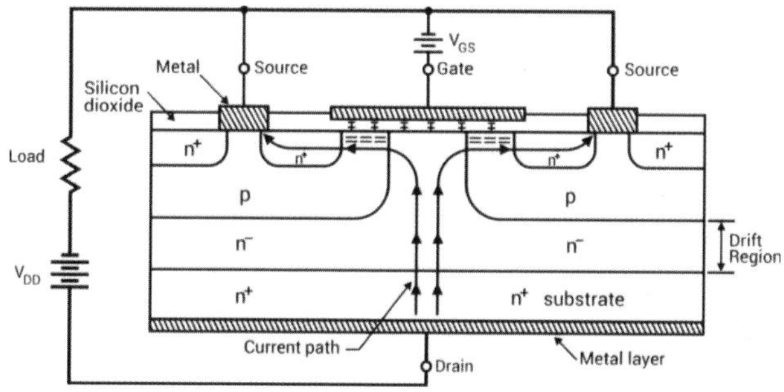

Figure: Power MOSFET Structural View with Connections

Drift region shown in Figure determines the voltage-blocking capability of the MOSFET.

When, $V_{GS} = 0$,

⇒ V_{DD} makes it reverse biased and no current flows from drain to source.

When, $V_{GS} > 0$,

⇒ Electrons form the current path as shown in figure. Thus, current from the drain to the source flows. Now, if we will increase the gate-to-source voltage, drain current will also increase.

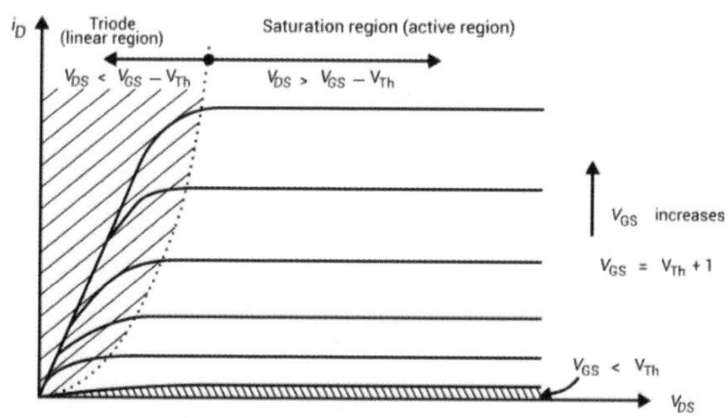

Figure: Drain Current (I_D) vs Drain-to-Source Voltage (V_{DS}) Characteristics Curves

For lower value of V_{DS}, MOSFET works in a linear region where it has a constant resistance equal to V_{DS} / I_D. For a fixed value of V_{GS} and greater than threshold voltage V_{TH}, MOSFET enters a saturation region where the value of the drain current has a fixed value.

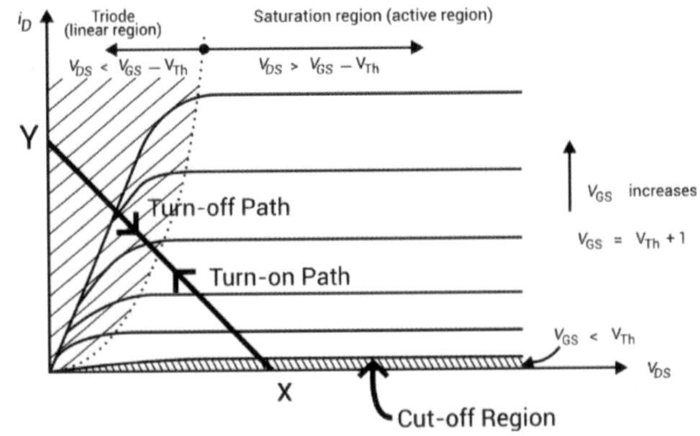

Figure: Output Characteristics with Load Line

If XY represents the load line, then the X-point represents the turn-off point and Y-point is the turn-on point where $V_{DS} = 0$ *(as voltage across the closed switch is zero).* The direction of turning on and turning off process is also shown in figure.

Besides the output characteristics curves, transfer characteristics of power MOSFET is also shown in figure.

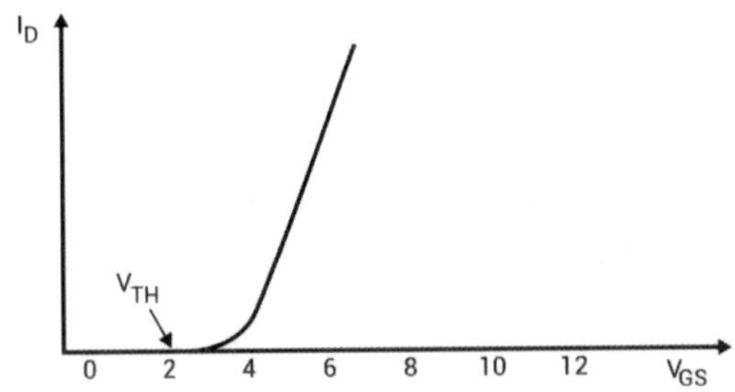

Figure: Gate-to-Source Voltage vs. Drain Current Characteristics for Power MOSFET

Here, V_{TH} is the minimum positive voltage between gate and the source above which MOSFET comes in on-state from the off-state. This is called threshold voltage. It is also shown in the output characteristics curve in figure.

Close view of the structural diagram given in figure reveals that there exists a fictitious BJT and a fictitious diode structure embedded in the power MOSFET as shown in figure.

As source is connected to both base and emitter of this parasitic BJT, emitter and base of the BJT are short circuited. That means this BJT acts in cut-off state.

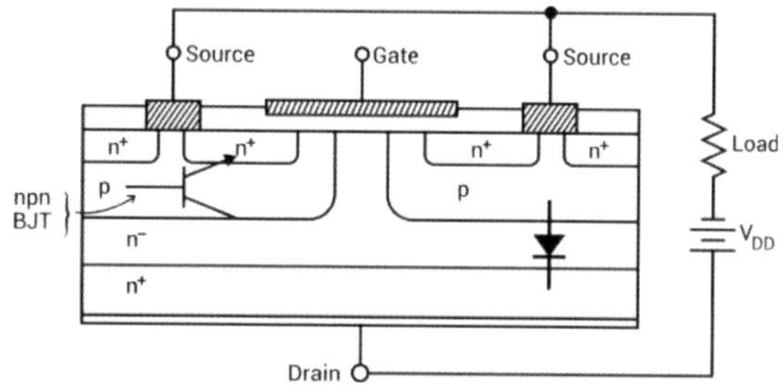

Figure: Fictitious BJT and Fictitious Diode in the Power MOSFET

Fictitious diode anode is connected to the source and its cathode is connected to the drain. So, if we apply the negative voltage V_{DD} across the drain and source, it will be forward biased. That means, the reverse-blocking capability of the MOSFET breaks. Thus, this can be used in inverter circuit for reactive loads without the need of excessive diode across a switch. Symbolically, it is represented in figure.

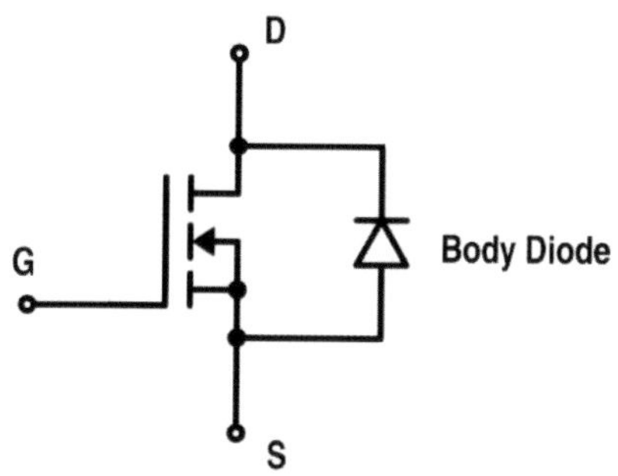

Figure: MOSFET Representation with Internal Body Diode

Although MOSFET internal body diode has sufficient current and switching speed for most of the applications, there may be some applications where the use of ultra-fast diodes is required. In such cases, an external fast-recovery diode is connected in an antiparallel manner. But a slow-recovery diode is also required to block the body diode action as given in figure.

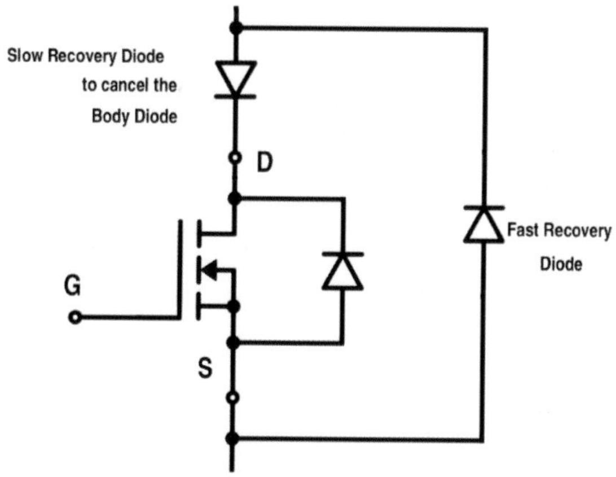

Figure: Implementation of Fast-Recovery Diode for Power MOSFET

One of the important parameters that affects the switching characteristics is the body capacitances existing between its three terminals i.e. drain, source and gate. Its representation is shown in figure.

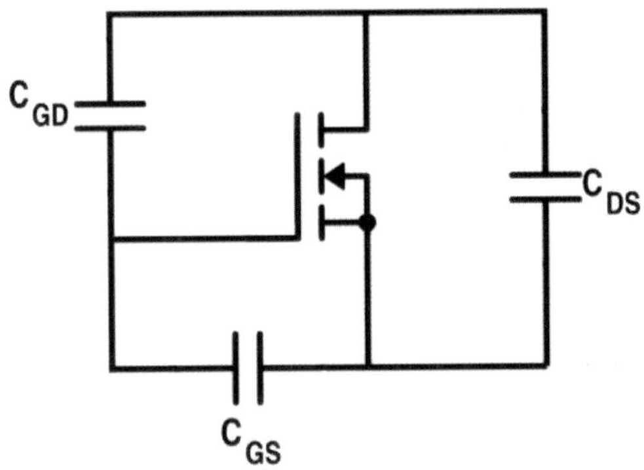

Figure: MOSFET Representation Showing Junction Capacitances

Parameters C_{GS}, C_{GD} and C_{DS} are all non-linear in nature and given in the device's data sheet of a particular MOSFET. They also depend on the DC bias voltage and the device's structure or geometry. They must be charged through gate during turn-on process to actually turn on the MOSFET. The drive must be capable of charging and discharging these capacitances to switch on or switch off the MOSFET.

Thus, the switching characteristics of a power MOSFET depend on these internal capacitances and the internal impedance of the gate drive circuits. Also, it depends on the delay due to the carrier transport through the drift region. Switching characteristics of power MOSFET are shown in figure.

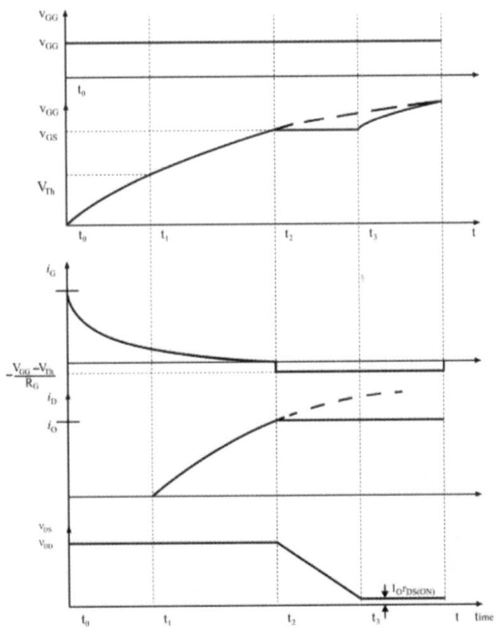

Figure: Turn-on Characteristics of Power MOSFET

There is a delay from t_0 to t_1 due to charging of input capacitance up to its threshold voltage V_{TH}. Drain current in this duration remains at zero value. This is called a delay time. There is a further delay from t_1 to t_2 during which the gate voltage rises to V_{GS}, a voltage required to drive the MOSFET into on-state. This is called the rise time. This total delay can be reduced by using a low-impedance drive circuit. The gate current during this duration decreases exponentially as shown. For the time greater than t_2, the drain current I_D has reached its maximum constant value I. As drain current has reached the constant value, the gate-to-source voltage is also constant as shown in the transfer characteristics of MOSFET in figure.

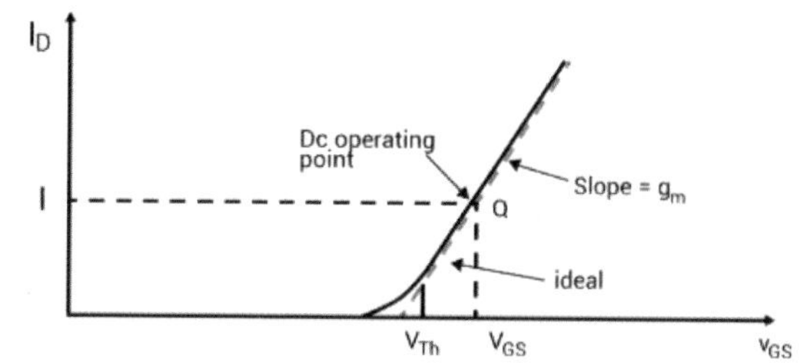

Figure: Transfer Characteristics of Power MOSFET with Operating Point

For turn-off characteristics, assume that the MOSFET is already in the switched-on situation with steady state. As $t = t_0$, gate voltage is reduced to zero value; C_{GS} and C_{GD} start to discharge through gate resistance R_G. This causes a turn-off delay time up to t_1 from t_0 as shown in figure. Assuming the drain-to-source voltage remains fixed. During this duration, both V_{GS} and I_G decreases in magnitude, drain current remains at a fixed value drawing current from C_{GD} and C_{GS}.

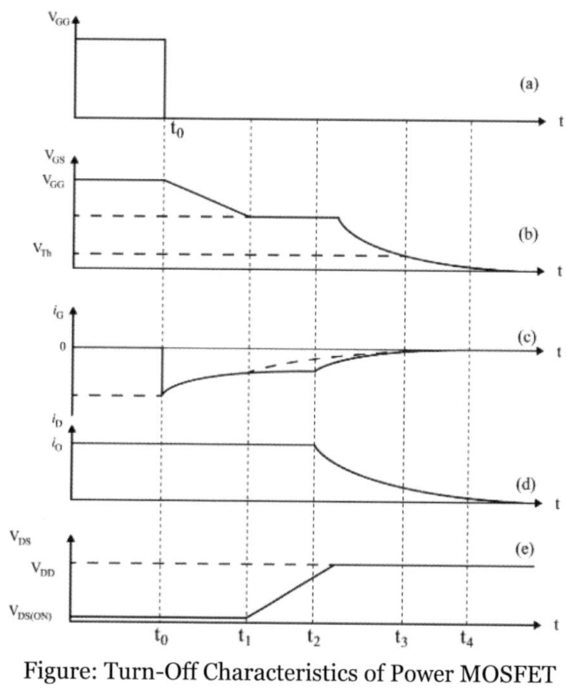

Figure: Turn-Off Characteristics of Power MOSFET

For the time where $t_2 > t > t_1$, gate-to-source voltage is constant. Thus, the entire current is now being drawn from C_{GD}. Up to time t_3, the drain current will almost reach zero value; which turns off the MOSFET. This time is known as the fall time, this is when the input capacitance discharges up to the threshold value. Beyond t_3, gate voltage decreases exponentially to zero until the gate current becomes zero.

Power Bipolar Junction Transistor (BJT)

Power BJT is used traditionally for many applications. However, IGBT (Insulated-Gate Bipolar Transistor) and MOSFET (Metal-Oxide-Semiconductor Field-Effect Transistor) have replaced it for most of the applications but still they are used in some areas due to its lower saturation voltage over the operating temperature range. IGBT and MOSFET have higher input capacitance as compared to BJT. Thus, in case of IGBT and MOSFET, drive circuit must be capable to charge and discharge the internal capacitances.

The BJT is a three-layer and two-junction npn or pnp semiconductor device as given in figure.

Although BJTs have lower input capacitance as compared to MOSFET or IGBT, BJTs are considerably slower in response due to low input impedance. BJTs use more silicon for the same drive performance.

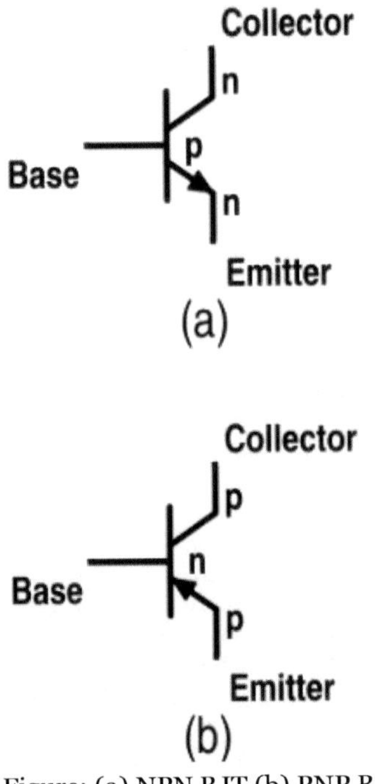

Figure: (a) NPN BJT (b) PNP BJT

In the case of MOSFET studied earlier, power BJT is different in configuration as compared to simple planar BJT. In planar BJT, collector and emitter is on the same side of the wafer while in power BJT it is on the opposite edges as shown in figure. This is done to increase the power-handling capability of BJT.

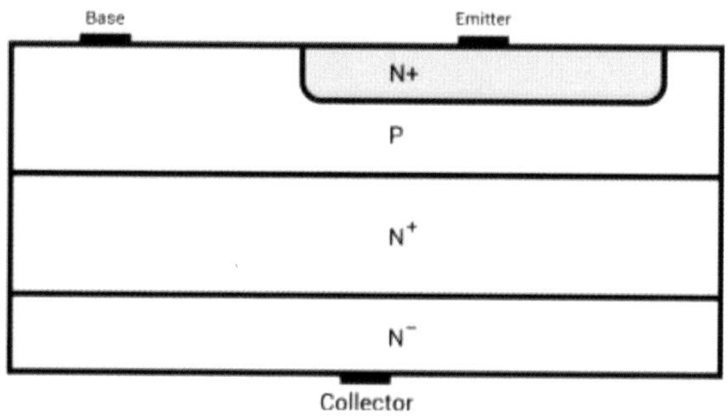

Figure: Power BJT PNP Structure

Power n-p-n transistors are widely used in high-voltage and high-current applications which will be discussed later.

Input and output characteristics of planar BJT for common-emitter configuration are shown in figure. These are current-voltage characteristics curves.

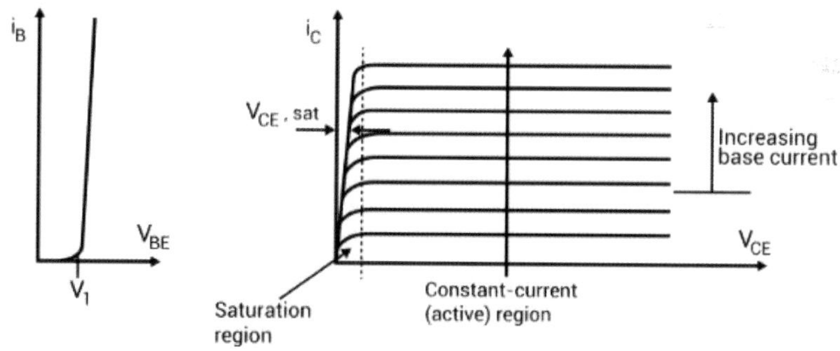

Figure: Input Characteristics and Output Characteristics for the Common-Emitter Configuration of Planar BJT respectively

Characteristic curves for power BJT is just the same except for the little difference in its saturation region. It has additional region of operation known as quasi-saturation as shown in figure.

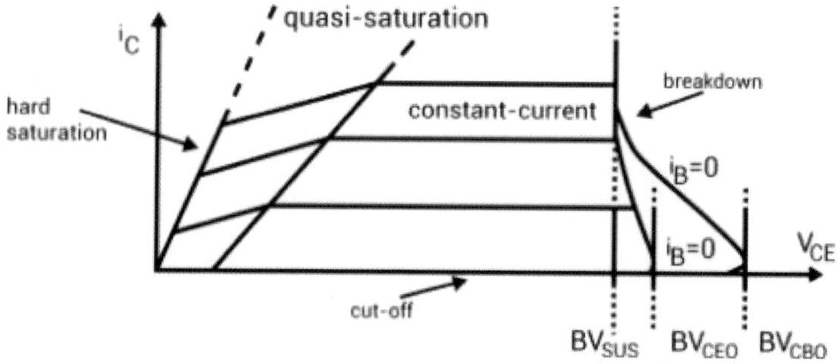

Figure: Power BJT Output Characteristics Curve

This region appears due to the insertion of lightly-doped collector drift region where the collector base junction has a low reverse bias. The resistivity of this drift region is dependent on the value of the base current. In the quasi-saturation region, the value of ß decreases significantly. This is due to the increased value of the collector current with increased temperature. But the base current still has the control over the collector current due to the resistance offered by the drift region. If the transistor enters in hard saturation region, base current has no control over the collector current due to the absence of the drift region and mainly depends on the load and the value of V_{CC}.

A forward-biased p-n junction has two capacitances named depletion layer capacitance and diffused capacitance. While a reverse bias junction has only a depletion capacitance in action. Value of these capacitances depends on the junction voltage and construction of the transistor. These capacitances come into role during the transient operation i.e. switching operations. Due to these capacitances, transistor does not turn on or turn off instantly.

Switching characteristics of power BJT is shown in figure. As the positive base voltage is applied, base current starts to flow but there is no collector current for some time. This time is known as the delay time(t_d) required to charge the junction capacitance of the base to emitter to 0.7 V approx. (known as forward-bias voltage). For $t > t_d$, collector current starts rising and V_{CE} starts to drop with the magnitude of 9/10th of its peak value. This time is called rise time, required to turn on the transistor. The transistor remains on so long as the collector current is at least of this value.

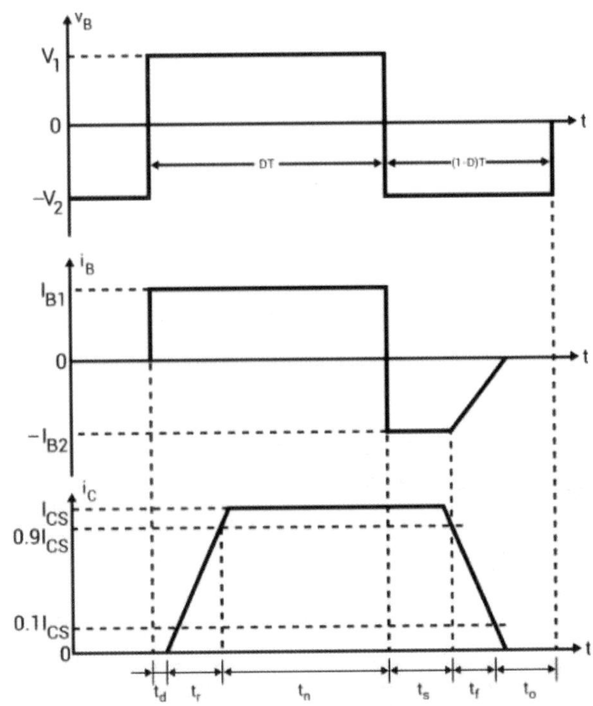

Figure: Turn-On and Turn-Off Characteristics of BJT

For turning off the BJT, polarity of the base voltage is reversed and thus the base current polarity will also be changed as shown in figure. The base current required during the steady-state operation is more than that required to saturate the transistor. Thus, excess minority carrier charges are stored in the base region which needs to be removed during the turn-off process. The time required to nullify this charge is the storage time, t_s. Collector current remains at the same value for this time. After this, collector current starts decreasing and base-to-emitter junction charges to the negative polarity; base current also get reduced.

Insulated-Gate Bipolar Transistor (IGBT)

IGBT combines the physics of both BJT and power MOSFET to gain the advantages of both worlds. It is controlled by the gate voltage. It has the high input impedance like a power MOSFET and has low on-state power loss as in case of BJT. There is no even secondary breakdown and not have long switching time as in case of BJT. It has better conduction characteristics as compared to MOSFET due to bipolar nature. It has no body diode as in case of MOSFET but this can be seen as an advantage to use external fast recovery diode for specific applications. They are replacing the MOSFET for most of the high voltage applications with less conduction losses. Its physical cross-sectional structural diagram and equivalent circuit diagram is presented in Figures. It has three terminals called collector, emitter and gate.

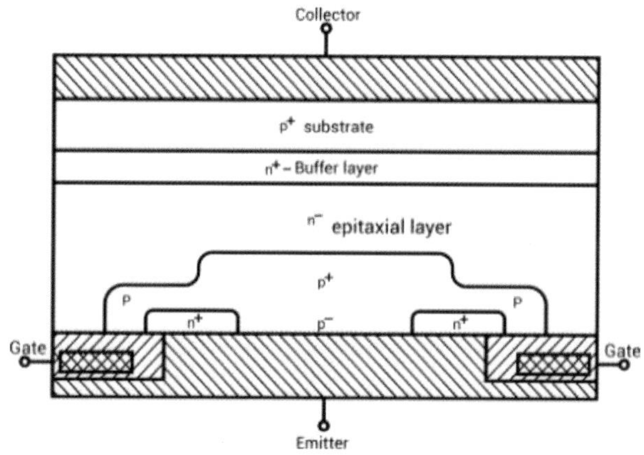

Figure: IGBT Structure View

There is a p+ substrate which is not present in the MOSFET and responsible for the minority carrier injection into the n-region. Gain of NPN terminal is reduced due to wide epitaxial base and n+ buffer layer.

There are two structures of IGBTs based on doping of buffer layer:

- Punch-through IGBT: Heavily doped n buffer layer → less switching time.
- Non-Punch-through IGBT: Lightly doped n buffer layer → greater carrier lifetime → increased conductivity of drift region → reduced on-state voltage drop.

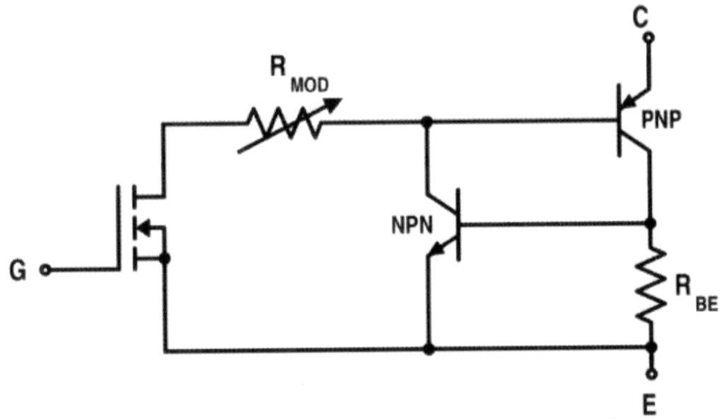

Figure: Equivalent Circuit for IGBT

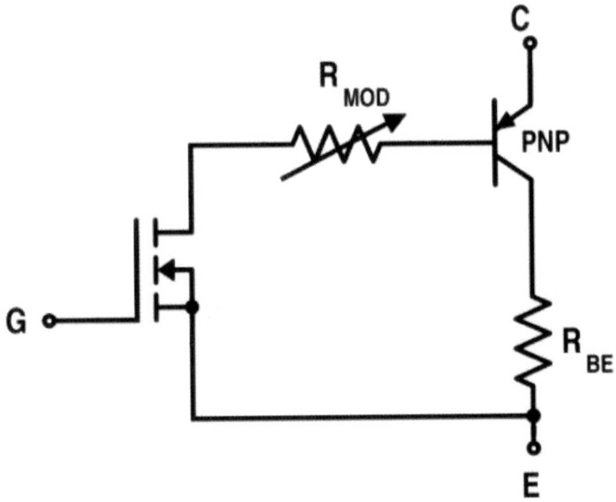

Figure: Simplified Equivalent Circuit for IGBT

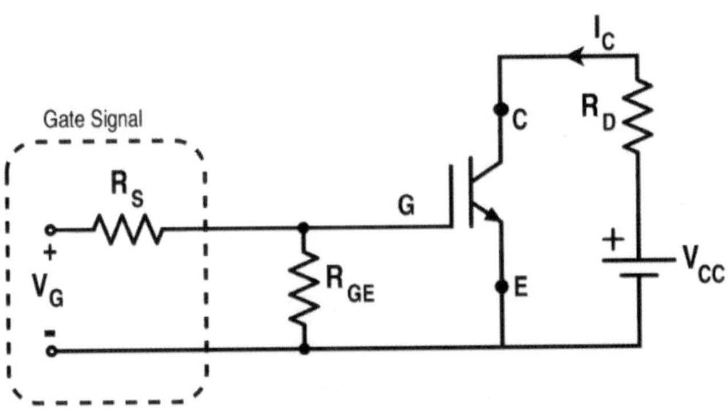

Figure: Circuit Diagram for IGBT

Based on this circuit diagram given in figure, forward characteristics and transfer characteristics are obtained which are given in figures. Its switching characteristic is also shown in figure.

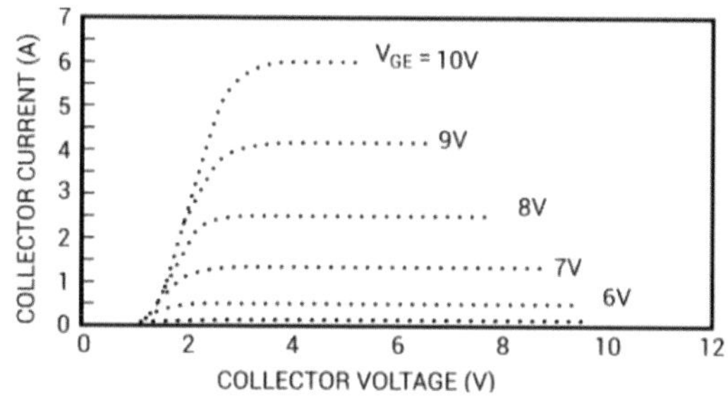

Figure: Forward Characteristics for IGBT

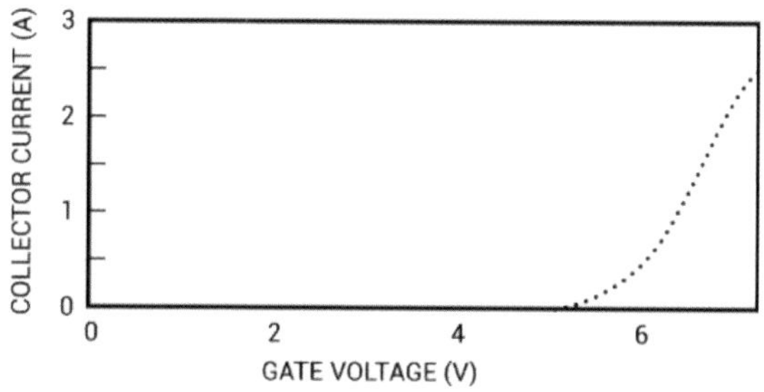

Figure: Transfer Characteristics of IGBT

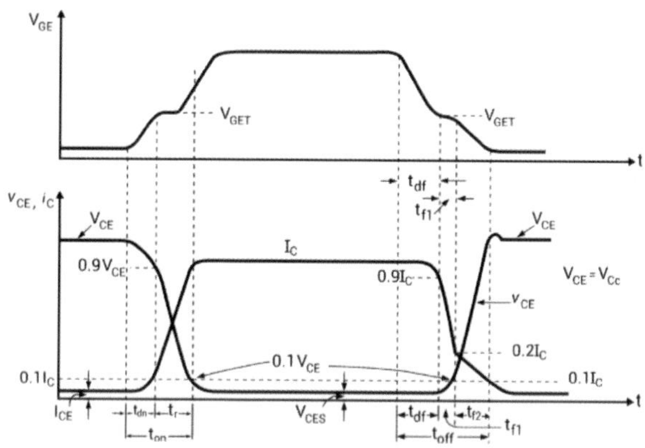

Figure: Turn-On and Turn-Off Characteristics of IGBT

(T_{dn} : delay time ; T_r: rise time ; T_{df}: delay time ; T_{f1}: initial fall time ; T_{f2}: final fall time).

Thyristors (SCR, GTO, MCT)

Thyristors are the family of solid-state devices extensively used in power electronics circuitry such as SCR (silicon-controlled rectifier), DIAC (diode on AC), TRIAC (triode on AC), GTO (gate turn-off thyristors), MCT (MOS-controlled thyristor), RCT, PUT, UJT, LASCR, LASCS, SIT, SITh, SIS, SBS, SUS, SBS and etc. SCR is the oldest member and the head of this family; and usually referred with the name "thyristor".

They are operated as bistable switches that are either working in non-conducting or conducting state. Traditional thyristors are designed without gate-controlled turn-off capability in which the thyristor can come from conducting state to non-conducting state when only anode current falls below the holding current. While GTO is a type of thyristor that has a gate-controlled turn-off capability.

SCR

SCR usually has three terminals and four layers of alternating p and n-type materials as shown in figure. The structure of the thyristor can be split into two sections: npn and pnp transistors for simple analysis purposes as shown in figure. It has three terminals named as cathode, anode and gate.

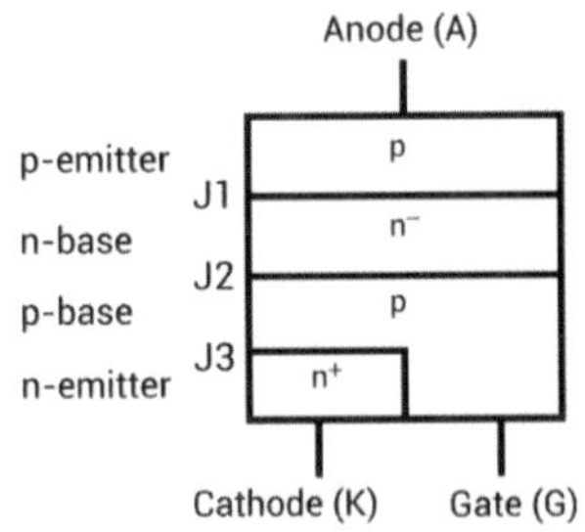

Figure: Structural View of Thyristor

N-base is a high-resistivity region and its thickness is directly dependent on the forward blocking rating of the thyristor. But more width of the n-base indicates a slow response time for switching. Symbol of thyristor is given in figure.

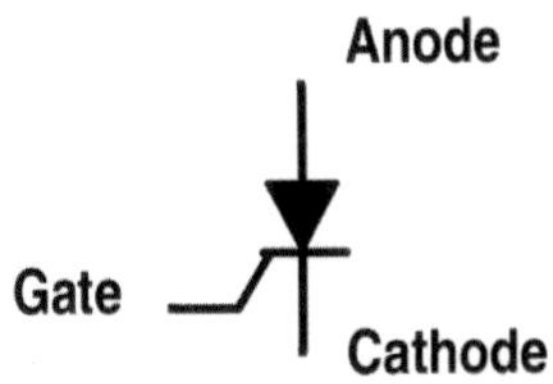

Figure: Schematic Symbol of Thyristor

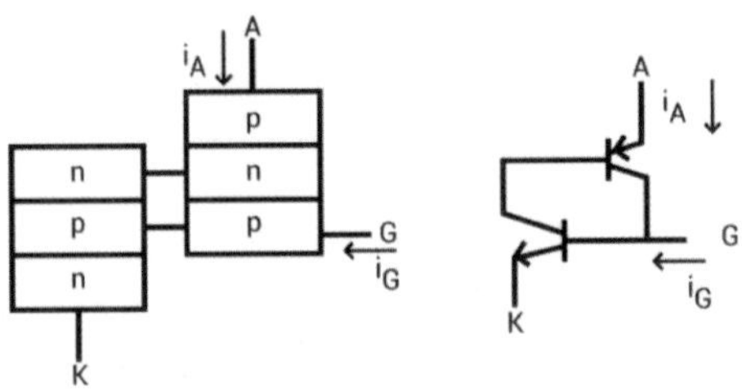

Figure: Two-Transistor Model of a Thyristor (A-Anode, G-Gate and K-Cathode)

Planar construction is used for low-power SCRs. In this type of construction, all the junctions are diffused. For high power, mesa construction is used where the inner layer is diffused and the two outer layers are alloyed on it.

The static characteristic obtained from the circuit given in below figure is drawn in below figure. It works under three modes: forward conducting mode, forward blocking mode and reverse blocking mode.

The minimum anode current that causes the device to stay at forward conduction mode as it switch from forward blocking mode is called the latching current. If the SCR is already conducting and the anode current is reduced from forward conducting mode to forward blocking mode, the minimum value of anode current to remain at the forward conducting mode is known as the holding current.

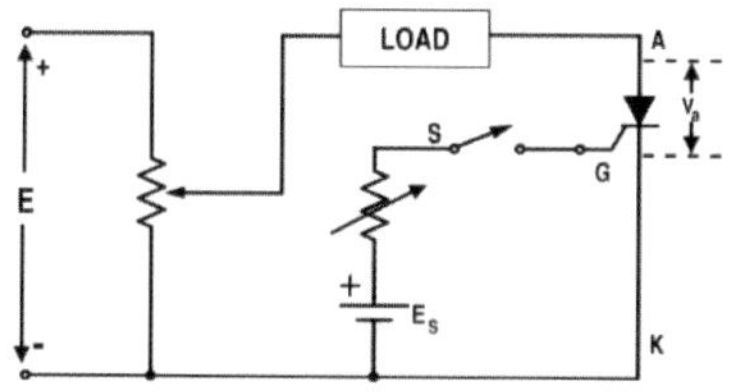

Figure: Basic Circuit for Getting Voltage and Current Characteristics of Thyristor

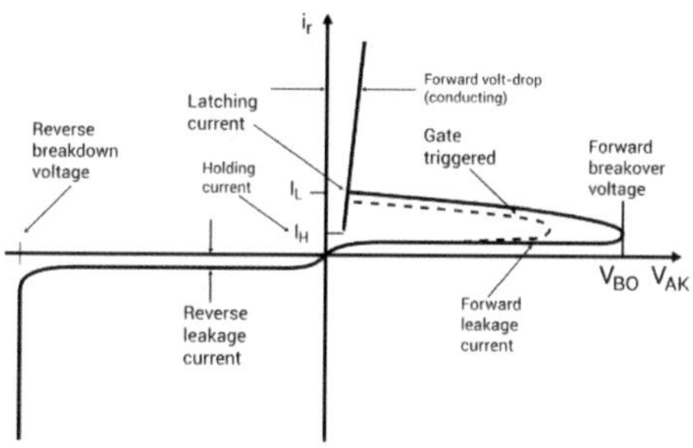

Figure: Static Characteristics Curve of SCR

Switching characteristics of SCR are shown in figure. Note that it can't be turned off with the gate. This is due to positive feedback or a regenerative feedback effect.

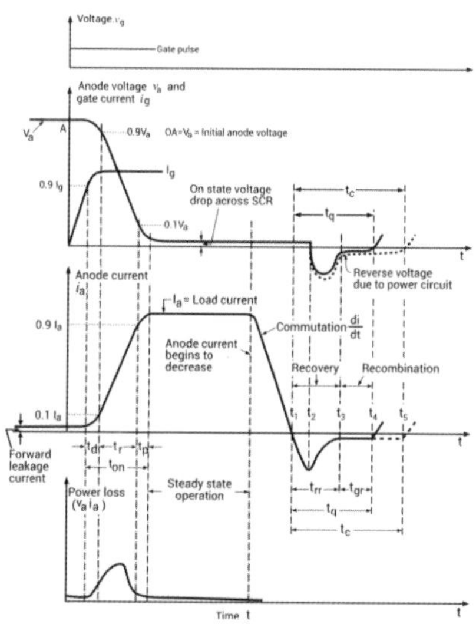

Figure: Turn-On and Turn-Off Characteristics of SCR

GTO (Gate Turn-off Thyristor)

GTO can be turned on with the positive gate current pulse and turned off with the negative gate current pulse. Its capability to turn off is due to the diversion of PNP collector current by the gate and thus breaking the regenerative feedback effect.

Actually the design of GTO is made in such a way that the pnp current gain of GTO is reduced. Highly doped n spots in the anode p layer form a shorted emitter effect and ultimately decreases the current gain of GTO for lower current regeneration and also the reverse voltage blocking capability. This reduction in reverse blocking capability can be improved by diffusing gold but this reduces the carrier lifetime. Moreover, it requires a special protection as shown in figure.

Figure shows the four Si layers, three junctions of GTO and its practical form:

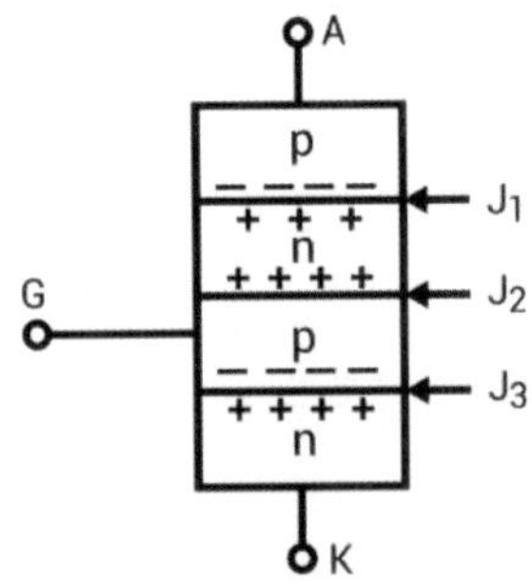

Figure: Four Layers and Three Junctions of GTO

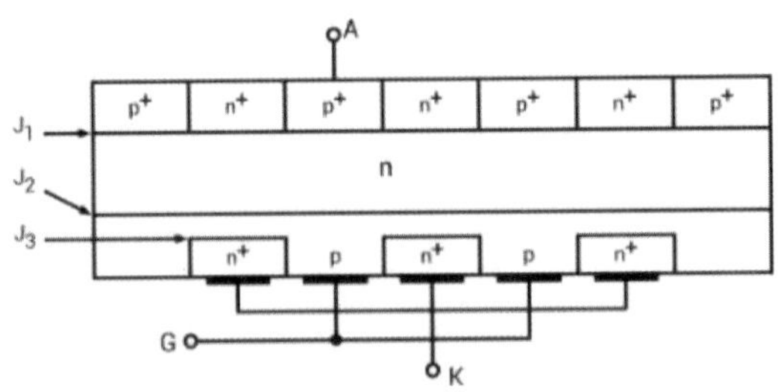

Figure: Practical Form of GTO

Overall switching speed of GTO is faster than thyristor (SCR) but voltage drop of GTO is larger. The power range of GTO is better than BJT, IGBT or SCR.

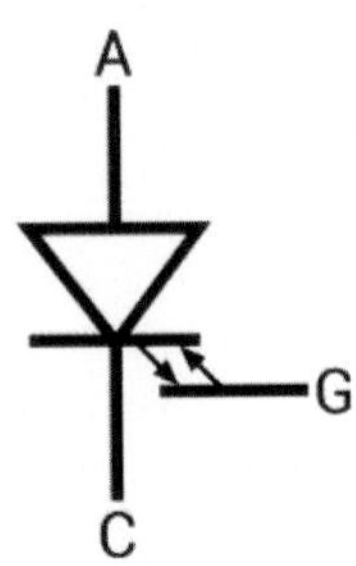

Figure: Symbol of GTO

The static voltage current characteristics of GTO are similar to SCR except that the latching current of GTO is larger (about 2 A) as compared to SCR (around 100-500 mA).

The gate drive circuitry with switching characteristics is given in figures.

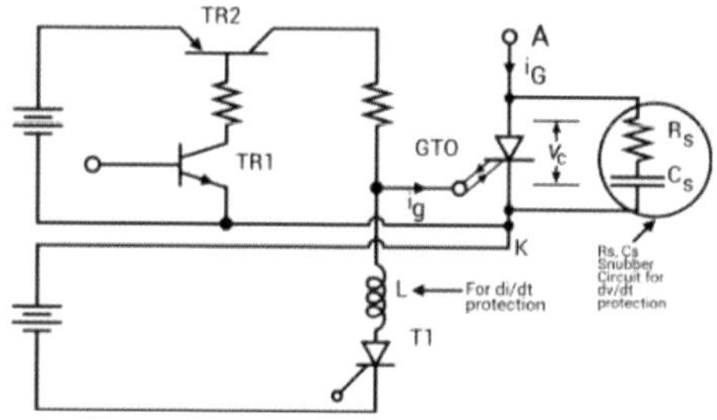

Figure: Gate Drive Circuit for GTO

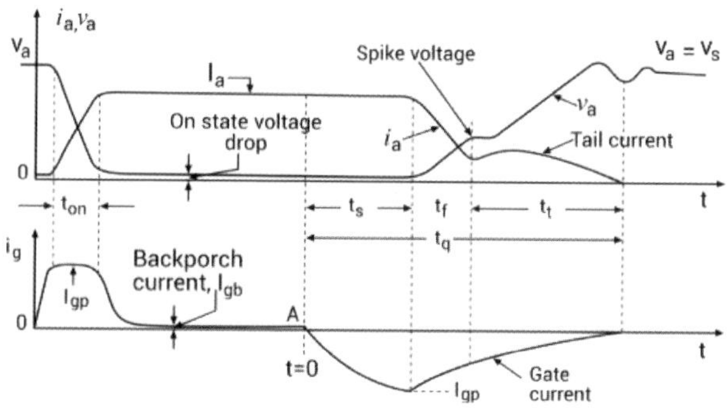

Figure: Turn-On and Turn-Off Characteristics of GTO

MCT (MOS-Controlled Thyristor)

IGBT is an improvement over a BJT using a MOSFET to switch on or switch off the anode current. Similarly, MCT is an improvement over a thyristor with a pair of MOSFETs to switch the current. There are several devices in the MCT's family but the p-channel is commonly discussed. Its schematic diagram and equivalent circuit is given in figures.

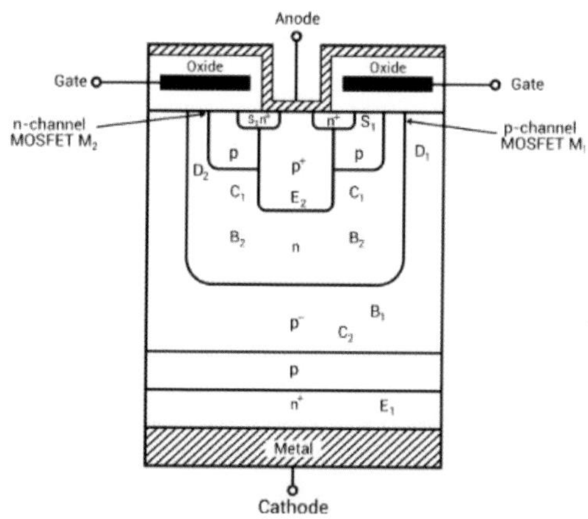

Figure: Schematic Diagram of P-Type MCT

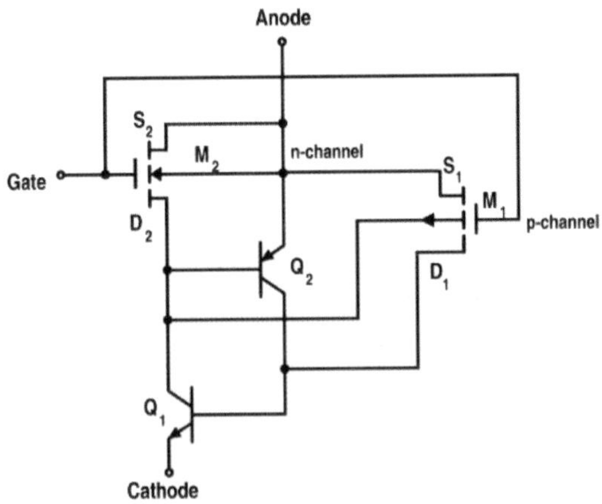

Figure: Equivalent Circuit for P-Type MCT

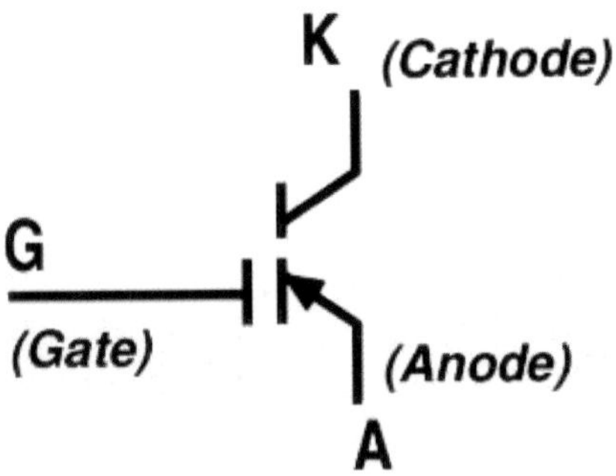

Figure: Symbol of P-Type MCT

Due to NPNP structure instead of PNPN, anode acts as a reference for gate. NPNP structure is represented by NPN transistor Q_1 and a PNP transistor Q_2 in the equivalent circuit.

The power required to switch it on or off is small with low switching losses due to its distributed structure across the entire surface of the device. Delay time due to charge storage is also small. It also has a low on-state voltage drop.

When a p-type MCT is in the forward-blocking state, it can be switched on by applying a negative pulse to its gate (with respect to anode). While when an n-channel MCT is in the forward-blocking mode, it can be switched on with the positive gate pulse (with respect to cathode). It will remain on until the device current is reversed or a turn-off pulse is applied to the gate i.e. applying a positive gate pulse for p-type MCT with respect to anode.

This device can bear a high current and high $\frac{di}{dt}$ capability. But just like any other devices, it needs to be protected against transient voltages and current spikes with the help of suitable snubbers. It is used in capacitor discharge applications, circuit breakers, AC-AC or AC-DC conversion. It is an ideal replacement for GTO as it requires a much simpler gate drive and certainly more efficient.

Power Electronic Converters

Modern power electronic converters are involved in a very broad spectrum of applications like switched-mode power supplies, active power filters, electrical-machine-motion-control, renewable energy conversion systems distributed power generation, flexible AC transmission systems, and vehicular technology, etc.

Power electronic converters can be found wherever there is a need to modify the electrical energy form with classical electronics in which electrical currents and voltage are used to carry information, whereas with power electronics, they carry power. Some examples of uses for power electronic systems are DC/DC converters used in many mobile devices, such as cell phones or PDAs, and AC/DC converters in computers and televisions. Large scale power electronics are used to control hundreds of megawatt of power flow across our nation.

Dual Converter

Dual converter is a combination of a rectifier and inverter in which the conversion of A.C to D.C happens and followed by D.C to A.C where load lies in between. A dual converter can be of a single phase or a three phase. A dual converter consists of two bridges consisting of thyristors in which one for rectifying purpose where alternating current is converted to direct current which can be given to load. Other bridge of thyristors is used for converting D.C to A.C.

Single Phase Dual Converter

Single phase dual converter uses a single phase as source which is given to converter 1 of dual converter for rectification followed to load.

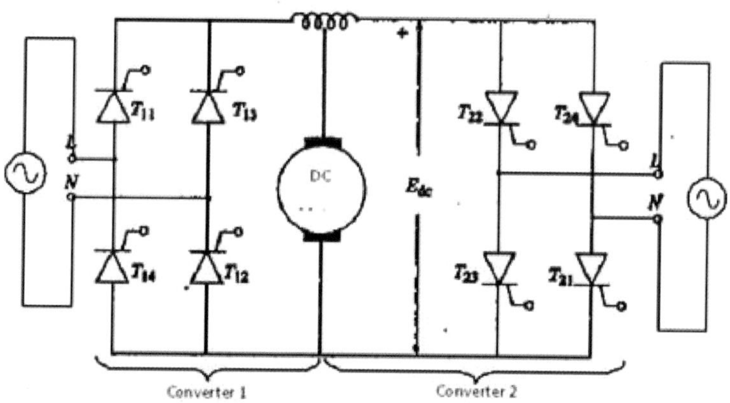

Principle of Operation

A.C input given to converter 1 for rectification in this process positive cycle of input is given to first set of forward biased thyristors which gives a rectified D.C on positive cycle, as well negative cycle is given to set of reverse biased thyristors which gives a D.C on negative cycle completing full wave rectified output can be given to load. During this process converter 2 is blocked using an inductor. As thyristor only start conducting when current pulse is given to gate and continuous conducting until supply of current is stopped. Output of Thyristor Bridge can be as follows when it is given to different loads.

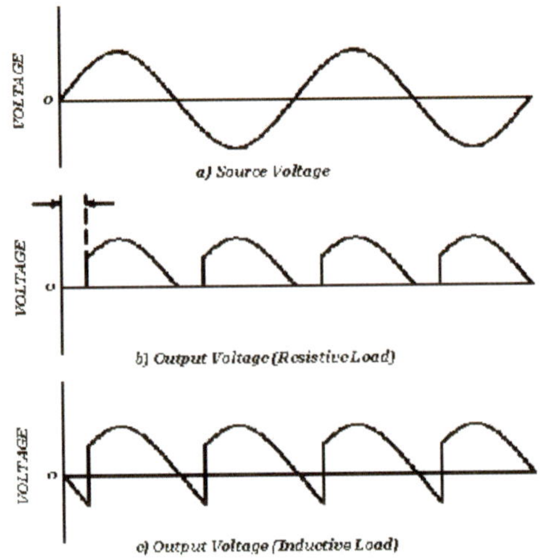

As a dual converter also consists conversion of D.C to A.C to make it work converter two is blocked, D.C inputs become load to dc power source conversion.

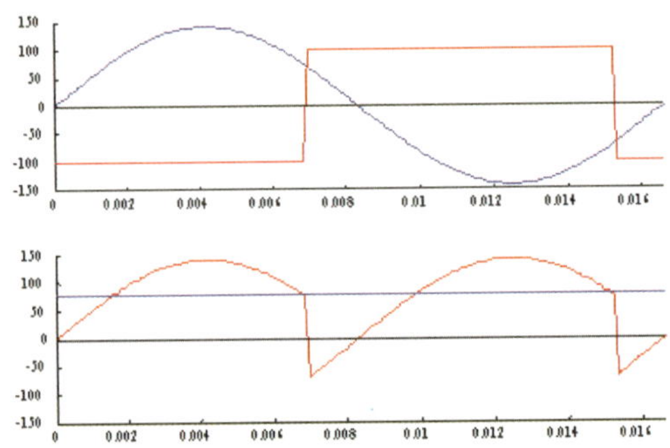

Firing of Thyristors

To make thyristors conduct, a trigger pulse must be given to its gate simultaneously along with line voltage. A separate gate drive circuit must be added to a dual converter thyristor bridges Gate drive circuit must be equally synchronized with source voltage, any delay causes zero cross jitter and zero frequency fluctuates. To prevent these circuits must be included with phase lock loops and comparators.

Applications of Single Phase Dual Converter

- Speed control and direction control in dc motors.

Speed Control and Polarity Control of DC Motor using Single Phase Dual Converter

A single phase dual converter can be used in controlling speed and direction of rotation interfacing with microcontroller, combination of four SCR's is placed either side of motor and motor is load. These thyristors can be triggered through an optocoupler which is connected to a port of microcontroller.

Rotation of motor can be initialized using optocoupler by setting a set of thyristor to trigger which is placed at one side and change in direction of motor can be achieved by triggering another set of thyristor Variation in speed of motor can be achieved by delayed firing angle of SCR.

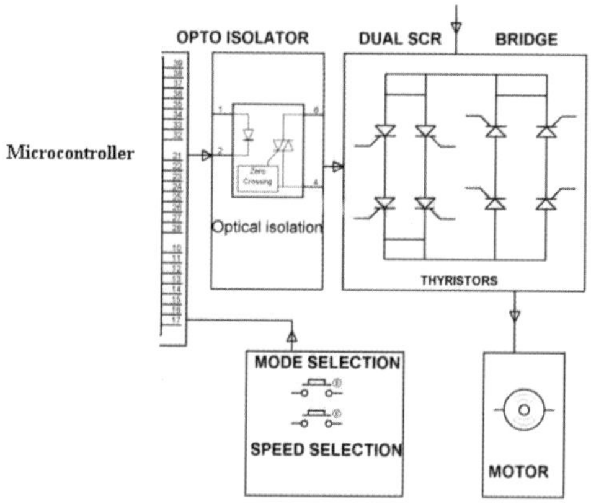

Mode selection and speed selection are microcontroller interfaced switches using these switches speed and rotation can be selected.

Single Phase Three-Leg AC/AC Converter

Power electronics is the application of electronics for power conversion. A subcategory of power conversion is the AC to AC conversion. An AC to AC voltage controller is a converter which controls the voltage, current and average power delivered to an AC load from an AC source. There are two types of AC voltage controllers, single and three phase AC controller:

A single phase AC/AC converter is a converter which converts from a fixed AC input voltage into variable AC output voltage with a desired frequency. They are used in practical circuits like light dimmer circuits, speed controls of induction motors and traction motor control etc. There are many existing technologies in single phase AC/AC converters; they are single phase – two legs, three legs and four legs. The single phase – two and four legs converters have some demerits like – they need large number of power devices, large control circuitry, more switching and losses are reduced only half to control the 50% of the output. So, to overcome these demerits present in the conventionally used converters, a better approach is use of single phase-three AC/AC converter.

A single phase – three legs consists of 3 legs and 6 switches. A leg is common for both grid side and load side. A leg performs the rectifier operation and a grid performs the inverter operation. And in this, we use Pulse Width Modulation(PWM) techniques for controlling the converter output.

During the positive half cycle of the supply voltage switches Qg and Qa in rectifier conducts and we get rectified output across the capacitor and for inverter operation in addition to the switches Qg and Qa', switch Ql in load side leg also triggered and we get ac output across the load. During negative half cycle switches Qa and Qg' in grid side conducts implying rectified output and for inversion operation in addition to the switches Qa and Qg', switch Ql' also triggered and we get ac output across the load. By using PWM method a fixed dc input voltage is supplied to the inverter and a controlled ac output voltage is obtained by adjusting the on and off periods of the inverter devices. The switches in the converter circuit for getting proper operation and also for reducing the harmonics. By varying the value of modulation index we can change the pulse width according to our convenience.

Advantages and Applications of 3 – Leg Converter

- The DC output voltage across the capacitor is almost doubled compared to the four leg converter.
- The power rating and voltage of the circuit can be improved.
- Same output can be obtained with reduced losses & switches. Hence the efficiency and the power factor can be improved.
- This converter is used in uninterruptable power supply circuits (UPS) and in power electronic for getting four quadrant operations of the drives.

Application of Power Electronics

Below is an attempt to briefly present the diaspora of power electronics.

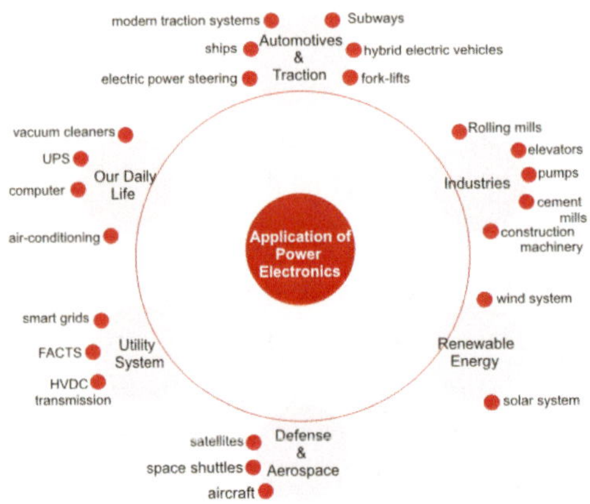

- Our Daily Life: If we look around ourselves, we can find a whole lot of power electronics applications such as a fan regulator, light dimmer, air-conditioning, induction cooking, emergency lights, personal computers, vacuum cleaners, UPS (uninterrupted power system), battery charges, etc.
- Automotives and Traction: Subways, hybrid electric vehicles, trolley, fork-lifts, and many more. A modern car itself has so many components where power electronic is used such as ignition switch, windshield wiper control, adaptive front lighting, interior lighting, electric power steering and so on. Besides power electronics are extensively used in modern traction systems and ships.
- Industries: Almost all the motors employed in the industries are controlled by power electronic drives, for eg. Rolling mills, textile mills, cement mills, compressors, pumps, fans, blowers, elevators, rotary kilns etc. Other applications include welding, arc furnace, cranes, heating applications, emergency power systems, construction machinery, excavators etc.
- Defense and Aerospace: Power supplies in aircraft, satellites, space shuttles, advance control in missiles, unmanned vehicles and other defense equipments.
- Renewable Energy: Generation systems such as solar, wind etc. needs power conditioning systems, storage systems and conversion systems in order to become usable. For example solar cells generate DC power and for general application we need AC power and hence power electronic converter is used.
- Utility System: HVDC transmission, VAR compensation (SVC), static circuit breakers, generator excitation systems, FACTS, smart grids, etc.

Optoelectronics

Optoelectronics is the communication between optics and electronics which includes the study, design and manufacture of a hardware device that converts electrical energy into light and light into energy through semiconductors. This device is made from solid crystalline materials which are lighter than metals and heavier than insulators. Optoelectronics device is basically an electronic device involving light. This device can be found in many optoelectronics applications like military services, telecommunications, automatic access control systems and medical equipments.

This academic field covers a wide range of devices including LEDs and elements, image pick up devices, information displays, optical communication systems, optical storages and remote sensing systems, etc. Examples of optoelectronic devices include telecommunication laser, blue laser, optical fiber, LED traffic lights, photo diodes and solar cells. Majority of the optoelectronic devices (direct conversion between electrons and photons) are LEDs, laser diodes, photo diodes and solar cells.

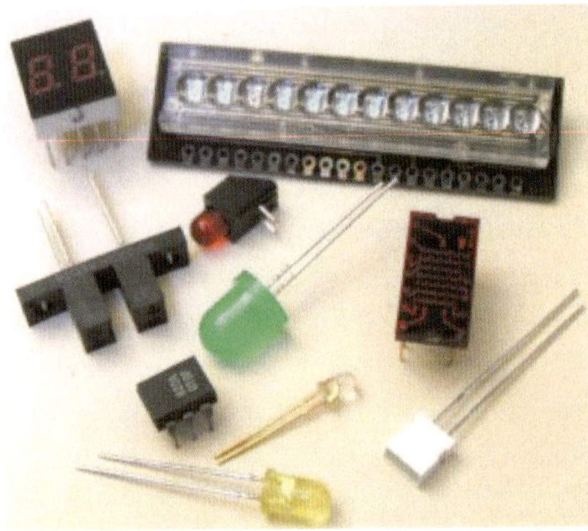

Optoelectronics Devices

Optoelectronic Devices

- Light absorption and emission
- Photodiodes
- Solar cells
- LEDs
- Laser diodes

P-n junctions are an integral part of several optoelectronic devices. These include photodiodes, solar cells light emitting diodes (LEDs) and semiconductor lasers. In this topic, we discuss the principle of operation of these devices and derive equations for key parameters.

Light Absorption and Emission

A large number of optoelectronic devices consist of a *p*-type and *n*-type region, just like a regular p-n diode. The key difference is that there is an additional interaction between the electrons and holes in the semiconductor and light. This interaction is not restricted to optoelectronic devices. Regular diodes are also known to be light sensitive and in some cases also emit light. The key difference is that optoelectronic devices such as photodiodes, solar cells, LEDs and laser diodes are specifically designed to optimize the light absorption and emission, resulting in a high conversion efficiency.

Light absorption and emission in a semiconductor is known to be heavily dependent on the detailed band structure of the semiconductor. Direct bandgap semiconductors, i.e semiconductors for which the minimum of the conduction band occurs at the same wave vector, k, as the maximum of the valence band, have a stronger absorption of light as characterized by a larger absorption coefficient. They are also the favored semiconductors when fabricating light emitting devices. Indirect bandgap semiconductors, i.e. semiconductors for which the minimum of the conduction band does not occur at the same wavevector as the maximum of the valence band, are known to have a smaller absorption coefficient and are rarely used in light emitting devices.

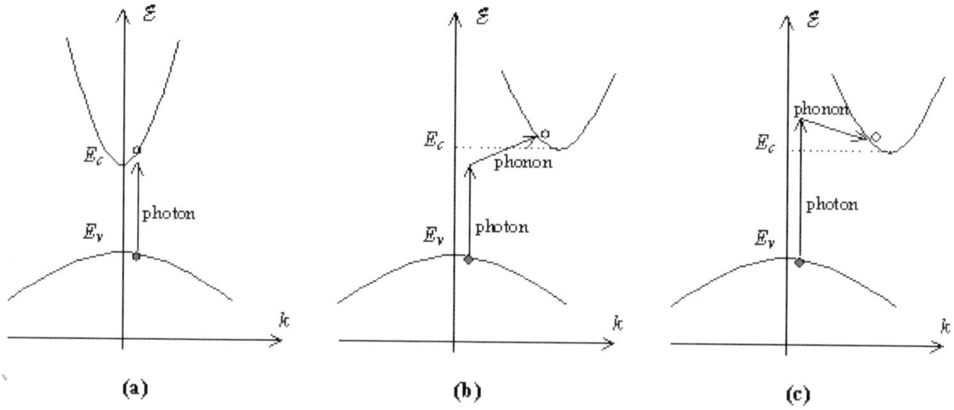

Figure: *E-k* diagram illustrating a) Photon absorption in a direct bandgap semiconductor b) Photon absorption in an indirect bandgap semiconductor assisted by phonon absorption and c) Photon absorption in an indirect bandgap semiconductor assisted by phonon emission.

This striking difference is further illustrated with Figure and can be explained based on the energy and momentum conservation required in the electron-photon interaction. The direct bandgap semiconductor, which has a vertically aligned conduction and valence band, is shown in figure (a). Absorption of a photon is obtained if an empty state in the conduction band is available for which the energy and momentum equals that of an electron in the valence band plus that of the incident photon. Photons have little momentum relative of their energy since they travel at the speed of light. The electron therefore makes an almost vertical transition on the E-k diagram.

For an indirect bandgap semiconductor, the conduction band is not vertically aligned to the valence band as shown in figure (b). Therefore a simply interaction of an incident photon with an electron in the valence band will not provide the correct energy and momentum corresponding to that of an empty state in the conduction band. As a result absorption of light requires the help of another particle, namely a photon. Since a phonon, i.e a particle associated with lattice vibrations, has a relatively low velocity close to the speed of sound in the material, it has a small energy and large momentum compared to that of a photon. Conservation of both energy and momentum can therefore be obtained in the absorption process if a phonon is created or an existing phonon participates. The phonon assisted absorption processes are illustrated with figure (b) and (c). Figure (b) illustrates the absorption of a photon aided by the simultaneous absorption of a phonon, while figure (c) depicts the absorption of a photon, which results in the emission of a phonon. The minimum photon energy that can be absorbed is slightly below the bandgap energy in the case of phonon absorption and has to be slightly above the bandgap energy in the case of phonon emission. Since the absorption process in an indirect bandgap semiconductor involves a phonon in addition to the electron and photon, the probability of having an interaction take place involving all three particles will be lower than a simple electron-photon interaction in a direct bandgap semiconductor. As a result one finds that absorption is much stronger in a direct bandgap material.

Similarly, in the case of light emission, a direct bandgap material is also more likely to emit a photon than an indirect bandgap material. While indirect bandgap materials are occasionally used for some LEDs, they result in a low conversion efficiency. Direct bandgap materials are used exclusively for semiconductor laser diodes.

Photodiodes

Photodiodes and crystalline solar cells are essentially the same as the p-n diodes, which have been described in this chapter. However, the diode is exposed to light, which yields a photocurrent in addition to the diode current so that the total diode current is given by:

$$I = I_s\left(e^{V_a/V_t} - 1\right) - 1_{ph}$$

Where, the additional photocurrent, I_{ph}, is due to photogeneration of electrons and holes shown in figure.These electrons and holes are pulled into the region where they are majority carriers by the electric field in the depletion region.

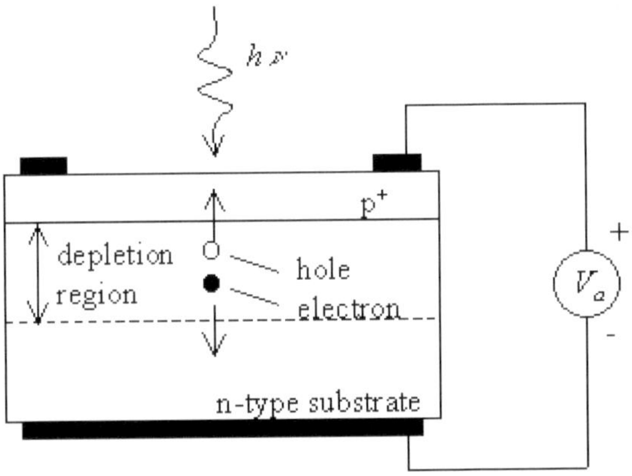

Figure: Motion of photo-generated carriers in a p-n photodiode.

The photo-generated carriers cause a photocurrent, which opposes the diode current under forward bias. Therefore, the diode can be used as a photodetector - using a reverse or even zero bias voltage - as the measured photocurrent is proportional to the incident light intensity. The diode can also be used as a solar cell - using a forward bias – to generate electrical power.

The primary characteristics of a photodiode are the responsivity, the dark current and the bandwidth. The responsivity is the photocurrent divided by the incident optical power. The maximum photocurrent in a photodiode equals:

$$I_{ph,max} = \frac{q}{h\nu} P_{in}$$

Where, P_{in} is the incident optical power. This maximum photocurrent occurs when each incoming photon creates one electron-hole pair, which contributes to the photocurrent. The maximum photocurrent in the presence of a reflection, *R* at the surface of the photodiode and an absorption over a thickness *d*, in a material with an absorption coefficient, α, is given by:

$$I_{ph} = (1-R)(1-e^{-\alpha d})\frac{qP_{in}}{h\nu}$$

This photocurrent is obtained by integrating the generation rate over the absorption region with thickness, *d*. The photocurrent is further reduced if photo-generated electron-hole pairs recombine within the photodiode instead of being swept into the regions where they are majority carriers.

The dark current is the current through the diode in the absence of light. This current is due to the ideal diode current, the generation/recombination of carriers in the depletion region and any surface leakage, which occurs in the diode. The dark current obviously limits the minimum power detected by the photodiode, since a photocurrent much smaller than the dark current would be hard to measure.

However, the true limitation is the shot noise generated by the current through the diode. The shot noise as quantified by the average of the square of the noise current is given by:

$$\langle i^2 \rangle = 2qI\Delta f$$

Where, I is the diode current and Df is the bandwidth of the detector. The bandwidth of the diode is affected by the transit time of the photo-generated carriers through the diode and by the capacitance of the diode. The carrier transit time yields the intrinsic bandwidth of the diode while the capacitance together with the impedance of the amplifier or the transmission line connected to the diode yields a the parasitic RC delay.

Solar Cells

Solar cells are typically illuminated with sunlight and are intended to convert the solar energy into electrical energy. The solar energy is in the form of electromagnetic radiation, more specifically "black-body" radiation. The sun's spectrum is consistent with that of a black body at a temperature of 5800 K. The radiation spectrum has a peak at 0.8 eV. A significant part of the spectrum is in the visible range of the spectrum (400 - 700 nm). The power density is approximately 100 mW/cm2.

Only part of the solar spectrum actually makes it to the earth's surface. Scattering and absorption in the earth's atmosphere, and the incident angle affect the incident power density. Therefore, the available power density depends on the time of the day, the season and the latitude of a specific location.

Of the solar light, which does reach a solar cell, only photons with energy larger than the energy bandgap of the semiconductor generate electron-hole pairs. In addition, one finds that the voltage across the solar cell at the point where it delivers its maximum power is less than the bandgap energy in electron volt. The overall power-conversion efficiency of single-crystalline solar cells ranges from 10 to 30 % yielding 10 to 30 mW/cm².

The calculation of the maximum power of a solar cell is illustrated by Figures. The sign convention of the current and voltage is shown as well. It considers a current coming out of the cell to be positive as it leads to electrical power generation. The power generated depends on the solar cell itself and the load connected to it. As an example, a resistive load is shown in the diagram below.

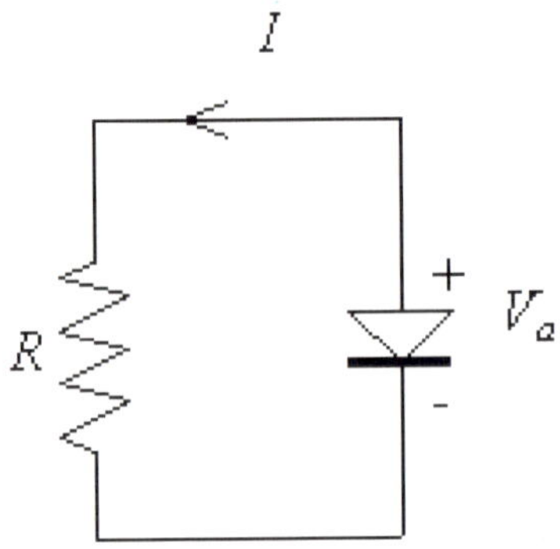

Figure: Circuit diagram and sign convention of a p-n diode solar cell connected to a resistive load.

The current and the power as function of the forward bias voltage across the diode are shown in figure for a photocurrent of 1 mA:

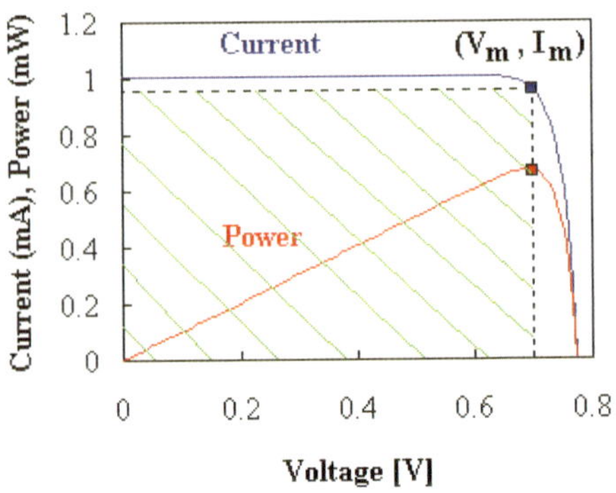

Figure: Current-Voltage (*I-V*) and Power-Voltage (*P-V*) characteristics of a p-n diode solar cell with I_{ph} = 1 mA and $I_s = 10^{-10}$ A. The crosshatched area indicates the power generated by the solar cell. The markers indicate the voltage and current, V_m and I_m, for which the maximum power, P_m is generated.

We identify the open-circuit voltage, V_{oc}, as the voltage across the illuminated cell at zero current. The *short-circuit* current, I_{sc}, is the current through the illuminated cell if the voltage across the cell is zero. The *short-circuit* current is close to the photocurrent while the *open-circuit* voltage is close to the turn-on voltage of the diode as measured on a current scale similar to that of the photocurrent.

The power equals the product of the diode voltage and current and at first increases linearly with the diode voltage but then rapidly goes to zero as the voltage approaches the turn-on voltage of the diode. The maximum power is obtained at a voltage labeled as V_m with I_m being the current at that voltage.

The fill factor of the solar cell is defined as the ratio of the maximum power of the cell to the product of the *open-circuit* voltage, V_{oc}, and the *short-circuit* current, I_{sc}:

$$\text{Fill Factor} = \frac{I_m V_m}{I_{sc} V_{oc}}$$

Example : A 1 cm² silicon solar cell has a saturation current of 10^{-12} A and is illuminated with sunlight yielding a short-circuit photocurrent of 25 mA. Calculate the solar cell efficiency and fill factor.

Solution: The maximum power is generated for:

$$\frac{dP}{dV_a} = 0 = I_s\left(e^{V_m/V_t}\right) - 1_{ph} + \frac{V_m}{V_t} I_s e^{V_m/V_t}$$

Where, the voltage, V_m, is the voltage corresponding to the maximum power point. This voltage is obtained by solving the following transcendental equation:

$$V_m = V_t \,\text{In} \frac{1 + I_{ph}/I_s}{1 + V_m/V_t}$$

Using iteration and a starting value of 0.5 V one obtains the following successive values for V_m:

$$V_m = 0.5,\ 0.542,\ 0.540 \text{ V}$$

and the efficiency equals:

$$\eta = \left|\frac{V_m I_m}{P_{in}}\right| = \frac{0.54 \times 0.024}{0.1} = 13\%$$

The current, I_m, corresponding to the voltage, V_m, was calculated using equation and the power of the sun was assumed 100 mW/ cm². The fill factor equals:

$$\text{fill factor} = \frac{V_m I_m}{V_{oc} I_{sc}} = \frac{0.54 \times 0.024}{0.62 \times 0.025} = 83\%$$

Where, the open circuit voltage is calculated using equation and $I = 0$. The short circuit current equals the photocurrent.

LEDs

Light emitting diodes are p-n diodes in which the recombination of electrons and holes yields a photon. This radiative recombination process occurs primarily in direct bandgap semiconductors where the lowest conduction band minimum and the highest valence band maximum occur at $k = 0$, where k is the wavenumber. Examples of direct bandgap semiconductors are GaAs, InP, and GaN while most group IV semiconductors including Si, Ge and SiC are indirect bandgap semiconductors.

The radiative recombination process is in competition with non-radiative recombination processes such as trap-assisted recombination. Radiative recombination dominates at high minority-carrier densities. Using a quantum well, a thin region with a lower bandgap, positioned at the metallurgical junction, one can obtain high carrier densities at low current densities. These quantum well LEDs have high internal quantum efficiency as almost every electron injected in the quantum well recombines with a hole and yields a photon.

The external quantum efficiency of planar LEDs is much lower than unity due to total internal reflection. As the photons are generated in the semiconductor, which has a high refractive index, only photons traveling normal to the semiconductor-air interface can exit the semiconductor. For GaAs with a refractive index of 3.5, the angle for total internal reflection equals 17° so that only a few percent of the generated photons can escape the semiconductor. This effect can be avoided by having a spherical semiconductor shape, which ensures that most photons travel normal to the interface. The external quantum efficiency can thereby be increased to values larger than 50%.

Laser Diodes

Laser diodes are very similar to LEDs since they also consist of a p-n diode with an active region where electrons and holes recombine resulting in light emission. However, a laser diode also contains an optical cavity where stimulated emission takes place. The laser cavity consists of a waveguide terminated on each end by a mirror. As an example, the structure of an edge-emitting laser diode is shown in figure. Photons, which are emitted into the waveguide, can travel back and forth in this waveguide provided they are reflected at the mirrors.

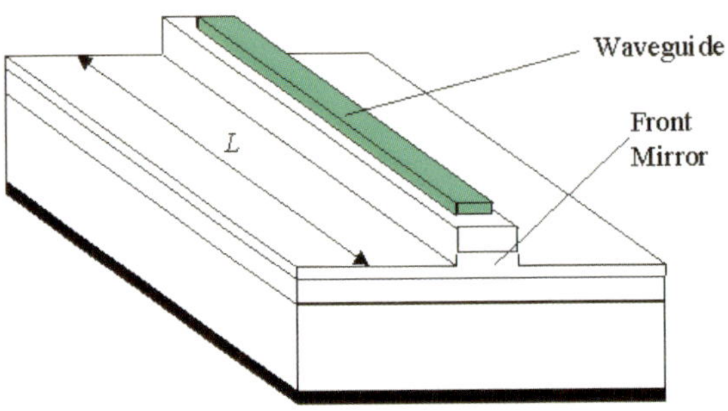

Figure: Structure of an edge-emitting laser diode.

The light in the waveguide is amplified by stimulated emission. Stimulated emission is a process where a photon triggers the radiative recombination of an electron and hole thereby creating an additional photon with the same energy and phase as the incident photon. This process is illustrated with figure. This "cloning" of photons results in a coherent beam.

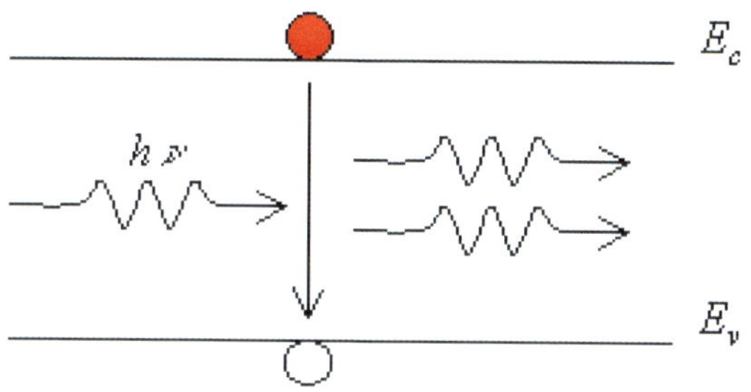

Figure: Stimulated emission of a photon.

The stimulated emission process yields an increase in photons as they travel along the waveguide. Combined with the waveguide losses, stimulated emission yields a net gain per unit length, g. The number of photons can therefore be maintained if the roundtrip amplification in a cavity of length, L, including the partial reflection at the mirrors with reflectivity R_1 and R_2 equals unity.

This yields the following lasing condition:

$$\text{Roundtrip amplification} = e^{2gL} R_1 R_2 = 1$$

If the roundtrip amplification is less than one, then the number of photons steadily decreases. If the roundtrip amplification is larger than one, the number of photons increases as the photons travel back and forth in the cavity. The gain required for lasing therefore equals:

$$g = \frac{1}{2L} \text{In} \frac{1}{R_1 R_2}$$

Initially, the gain is negative if no current is applied to the laser diode as absorption dominates in the waveguide. As the laser current is increased, the absorption first decreases and the gain increases. The current for which the gain satisfies the lasing condition is the threshold current of the laser, I_{th}. Below the threshold current very little light is emitted by the laser structure. For an applied current larger than the threshold current, the output power, P_{out}, increases linearly with the applied current, as each additional incoming electron-hole pair is converted into an additional photon. The output power therefore equals:

$$P_{out} = \eta \frac{h\nu}{q} (1 - 1_{th})$$

where $h\nu$ is the energy per photon. The factor, η, indicates that only a fraction of the generated photons contribute to the output power of the laser as photons are partially lost through the other mirror and throughout the waveguide.

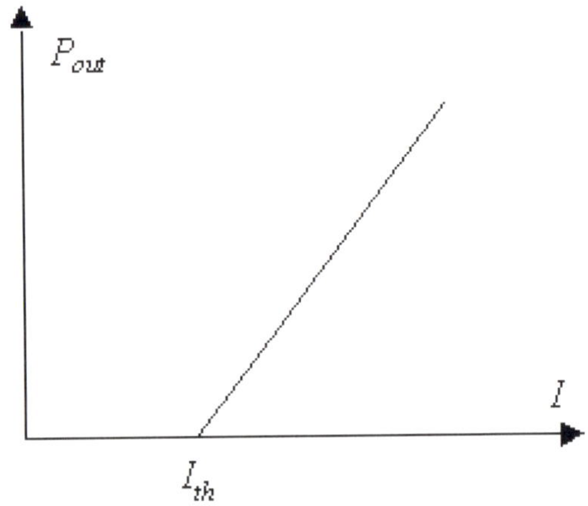

Figure: Output power from a laser diode versus the applied current.

Applications of Optoelectronics Devices

Mains Operated LED

- LEDs could become the next generation of lighting and used anywhere like in indication lights, computer components, medical devices, watches, instrument panels, switches, fiber-optic communication, consumer electronics, household appliances, traffic signals, automobile brake lights, 7 segment displays and in-active displays, and also used in different electronic and electrical engineering projects such as:

1. Propeller Display of Message by Virtual LEDs;
2. LED Based Automatic Emergency Light;
3. Mains Operated LED Light;

4. Display of Dialed Telephone Numbers on Seven Segment Display;
5. Solar Powered Led Street Light with Auto Intensity Control.

- The solar cells are applicable in rural electrification, telecommunication systems, ocean navigation aids, and electric power generation in space and remote monitoring and control systems and also used in different solar energy based projects such as:

1. Solar Energy Measurement System;
2. Arduino based Solar Street Light;
3. Solar Powered Auto Irrigation System;
4. Solar Power Charge Controller;
5. Sun Tracking Solar Panel.

- Photodiodes are used in many types of circuits and different applications such as cameras, medical instruments, safety equipments, industries, communication devices and industrial equipments.

Solar based Project

- The laser diodes are used in fiber optic communication, optical memories, military applications, CD players, surgical procedures, Local Area Networks, long distance communications, optical memories, fiber optic communications and in electrical projects such as RF Controlled Robotic Vehicle with Laser Beam Arrangement and so on.

Thus, this is all about the optoelectronic devices which include laser diodes, photo diodes, solar cells, LEDs, optical fibers.These optoelectronic devices are used in different electronic project kits as well as in telecommunications, military services and in medical applications.

Printed Electronics

Printed electronics is an all-encompassing term for the printing method used to create electronic devices by printing on a variety of substrates. Originally, printed electronics related to organic or plastic electronics that use one or more inks made of carbon-based compounds. As demand for wearable devices and thinner electronics expands, printed electronics are being used to form flexible keyboards, antennas, electronic skin patches, and more. Printed electronics technology has evolved over time, and now inkjet printers are capable of printing electrical circuits quite inexpensively and quickly. In short, printed electronics is one of the fastest growing technologies today and is becoming invaluable to several industries including healthcare, aerospace, media, and transit.

Printing Technologies

Enabling the use of roll-to-roll and high-throughput device fabrication technologies is the most relevant aspect of printed electronics. Throughputs greater than 1 m^2/s are considered "high volume"; most printing methods fall in this category. There are several considerations to determine what process can be used based mainly on the viscoelastic properties of materials and the desired feature sizes (lateral resolution, film thickness, surface morphology, surface energy) required for device assembly. Table shows some of the most important printing techniques and relevant specifications for use in electronics; probably the most used processes are inkjet, screen, and gravure printing.

Inkjet Printing

This process uses a stepper motor to control the position of the print head along a stabilizer bar; as the print head slides back and forth along the bar, ink drops are ejected from the nozzle to create a pattern on the substrate. There are two primary mechanisms for ejecting the ink: in a thermal inkjet, evaporation of a small portion of the ink solvent forces ink out of the nozzle; in a piezoelectric inkjet, voltage applied to a piezoelectric material causes it to expand, forcing ink out of the nozzle.

Table: Typical ink reauirements, printing features, and throughputs of conventional printing methods.

Technique	Ink Viscosity (mPas)	Film Thickness (μm)	Resolution (μm)	Registration (μm)	Throughput (m^2/s)	Features/ Issues	Electronic Materials Printed
Lithography	$(10–4) \times 10^2$	1.5–0.5	50–10	>10	0.1	High quality; Need for additives	Conductor
Screen	$(5–0.5) \times 10^2$	100–30	100–30	>25	2–3	Wide range of inks; Medium quality	Conductor

Flexography	5000–50	2.5–0.8	70	<200	10	Wide range of substrates; Medium quality	Conductor Semiconductor Dielectric
Gravure	200–50	5–0.5	<10	>5	60	Large run length; High quality	Dielectric Semiconductor
Pad	>50	2–1	20	>10	0.1	Nonplanar objects	Conductor
Laser/thermal	N/A	<1	5	~10	0.002	N/A	Conductor Semiconductor
Inkjet	30–1	<0.5	<10	20–5	0.5–0.01	Digital data; Local registration	Conductor Semiconductor
N/A = not applicable							

Inkjet printing is now one of the most widespread graphic arts printing methods, and in recent years has received attention as a technique to deposit functional materials with specific electrical, optical, chemical, biological, or structural functionalities at well-defined locations on a substrate.

One of the most useful features of inkjet printing is its capacity to vary the printing pattern without the need to make a new printing plate. Using a camera and image analysis software, the printed image can be adjusted "on the fly" to compensate for many of the registration errors that plague other types of printing process. Organic semiconductors are commonly inkjet patterned for transistor and OLED fabrication. Recently a variation on inkjet printing, self-aligned printing, was used to pattern features as small as 60 nm.

Inkjet printing has some limitations, such as susceptibility of the printing head to corrosion from aggressive solvents, a liability to high mechanical shears in piezoelectric print heads, and high temperatures in thermal inkjet heads. In addition, fluctuations in droplet volume or trajectory can adversely affect film uniformity (creating a "coffee-stain" effect) and materials performance.

Screen Printing

Screen printing consists of three key elements—the screen, which is the image carrier; the squeegee; and the ink—and uses a porous polyester or stainless steel mesh stretched tightly over a wood or metal frame. A stencil, produced on the screen either manually or photochemically, defines the image to be printed (in other printing technologies this is called an image plate). The technique usually produces relatively thick films (thinner films fabricated using less viscous inks often result in poorly defined printed patterns); the resolution is poor and limited by the screen size, although recent presses perform much better.

In organic electronics this technique has been used to fabricate top-level interconnects and contacts where the thickness of the printed film is not a critical factor. It can also be used to print thick dielectric layers and passive materials for device encapsulation. Indeed, the first report of a "printed organic thin film transistor" described screen-printed carbon paste electrodes for source, drain, and gate contacts.

Gravure Printing

In gravure printing the image areas consist of honeycomb-shaped cells that typically are etched or engraved on a copper cylinder. The cylinder rotates into an ink pan and, as it turns, excess ink is removed by a flexible steel doctor blade so that the ink remaining in the recessed cells forms the image by direct transfer to the substrate as it passes between the plate cylinder and the impression cylinder. The process uses fluid inks with relatively low viscosity.

Because it is one of the highest-volume printing processes, gravure printing is often used commercially to produce high-quality graphic materials such as magazines. Depending on the nature (depth, surface energy, shape, etc.) of the engraved cells, different amounts of material used to fabricate the devices (semiconductor, conductor, dielectric) can be deposited in different substrate regions, enabling tuned film thicknesses. The drawback is that the printed pattern edges may be wavy.

Gravure has recently been used to fabricate dielectric and semiconducting layers for organic transistors.

Printed Electronics Applications

Printed electronics are being used in more products as the technology continues to advance. Today, flexible screens, intelligent labels and packaging, interactive books and posters, and even upholstery are being created with printed electronics. As the substrates grow thinner, printed materials become thin, light, and flexible enough to be integrated into existing production lines.

The market is catching up to the technology, according to Dr. Harry Zervos, principal analyst and business development manager for North America for IDTechEx: "The market for printed and flexible electronics is set to reach over $73 billion by 2025. Not all separate types of printed electronics will share equally in that big market. Printed electronics is an umbrella term and some markets will reach maturity sooner than others – but already, OLEDs, conductive inks in PV applications and printed glucose sensors are billion dollar markets that are set to grow."

And, as Thin Film Electronics' chief strategy officer Jennifer Ernst points out, the market for printed electronics is growing because the Internet of Things is expanding and requires low-cost, lightweight technology that can sense, store information securely, and transmit data. For these reasons, printed memory, sensors, and communications devices will play a part in the IoT market.

Researchers continue to evolve printed electronics. For example, MIT engineers are working with a new type of printed electronics that involves an intricate printing process and stamping technique capable of printing transistors small enough to control individual pixels in high-resolution displays and touch screens. These engineers have succeeded in creating a stamp that prints electronic inks onto rigid or flexible surfaces. Their new printing technique delivers an inexpensive, fast way to manufacture electronic surfaces for applications that have yet to be created.

Benefits of Printed Electronics

Printed electronics have become secure, flexible, and cost-effective, all of which make them appealing to a broad range of industries. Printed circuitry has the potential to reduce costs and technical constraints typically associated with mass producing electronics. Printed electronics also require fewer input materials and less energy to work with them. And, printed electronics pave the way for flexible devices that people previously may not have thought possible. For example, companies are working on using printed electronics for identifying banknotes, credit cards, legal documents, and other items with unique printed signatures. Another industry benefiting from printed electronics advancements is photovoltaics. Printed electronics have the potential to significantly change solar power projects, thanks to less expensive polymer electronics.

Overall, the benefits of printed electronics include:

- Low cost
- Attractive and flexible form factor
- Ease of production
- Ease of integration

- Facilitating widespread development of non-conventional functional electronic devices including flexible displays, smart labels, animated posters, active clothing, and more.

Challenges with Printed Electronics

Despite its status as a fast-growing field in technology, printed electronics is not without its challenges. Companies and researchers continue to develop the best conductive inks for printed electronics and strive to find the materials that will make their printed electronics among the best available. Other challenges with printed electronics stem from the fact that the processes are so new that trial-and-error and testing are critical to success.

References

- Digital-electronics: electronicscoach.com, Retrieved 31 March 2018
- Introduction-analog-electronics: engineersgarage.com, Retrieved 11 May 2018
- A-review-on-power-semiconductor-devices, technical-articles: allaboutcircuits.com, Retrieved 31 March 2018
- Power-electronic-converters: elprocus.com, Retrieved 01 July 2018
- Optoelectronics-devices-with-their-applications: elprocus.com, Retrieved 28 May 2018
- What-is-printed-electronics: pannam.com, Retrieved 08 April 2018

Electrical Elements

Electrical elements are the components of an electrical circuit. These may refer to active and passive electrical components. Some of the important active elements are diodes, transistors and batteries, while passive elements may include resistors, capacitors and inductors. This chapter discusses in detail the working of these electrical elements and their significance in an electrical circuit.

Electrical Elements or Electronic components are the elements of circuit which helps in its functioning. They can be classified into two types i.e. Active Components and Passive Components. Active elements are those which possess gain. They can give energy to the circuit. On the contrary passive elements do not possess gain and they cannot give energy continuously to the circuit.

Passive components cannot amplify or energize the energy of the signal associated with them, they can only attenuate it, while active components can energize or amplify the signal. For example, transistors, amplifiers are active elements. While resistor, inductor, capacitor are passive elements as they cannot supply energy to the circuit.

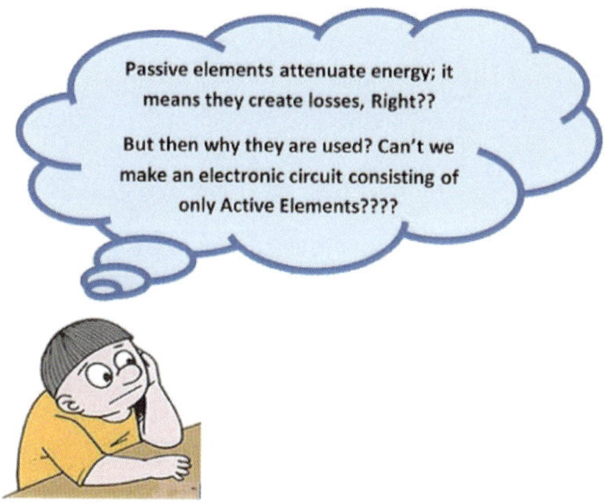

No, you cannot design a circuit or synthesize it before connecting a passive element in it. Passive elements holds significance in processing of various tasks.

To understand the importance of Resistor, let us consider a circuit consisting of diode and voltage source at input. If the current though the diode becomes excessive large then it may lead to the damage of the diode. The condition may become worse due to breakdown of circuit. Thus, it is very crucial to limit the value of current through the diode to protect it.

Another significant passive devices are inductor and capacitor which stores energy supplied by active device. If inductor and capacitor is not placed in a circuit, the energy flowing from active source will continue to flow and will cease to flow only when the device switches to off state. Thus, energy will not be stored anywhere in the circuit. Such a processing is totally futile if we cannot use the stored energy during the inactive state of the voltage source.

The capacitor stores energy in the form of electric field and the inductor stores energy in the form of magnetic field. When the active source is switched ON the capacitor starts get charging and get charged to its peak value. When the active source is switched off the capacitor will starts get discharging and provides energy to the circuit even when the active source is OFF.

Similarly, inductor also performs the same activity. But it stores energy in the form of magnetic field.

Classsification of Electronic Components

Electronic component can be defined into various types based on various parameter such as measurement components, power supply components etc.

- Elementary Components based on design architecture: Diodes, Transistors, resistors, capacitors, inductors, LED etc. These are the basic components of a circuit as without these components the circuit cannot not process or work properly.

- Power Supply Components: DC power supplies, AC power supply, signal generators are the components which fall in this category. These are also the considered as active components as they supply energy to the circuit.
- Measurement devices: Cathode ray oscilloscope and millimeters are usual measurement devices in electronic circuits.

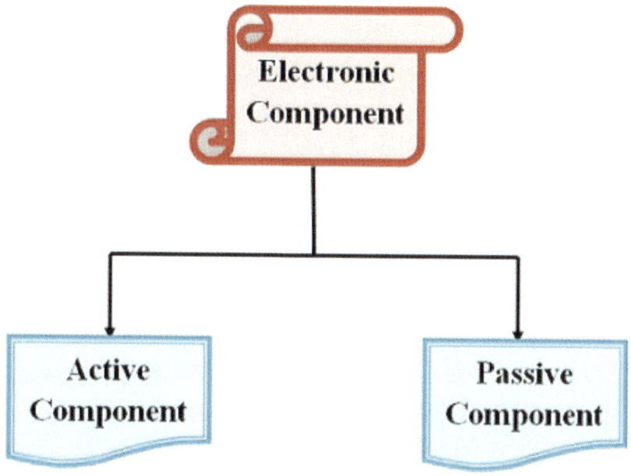

The above diagram shows the classification of electronic Components.

Active Components

Active devices supply energy to the circuit. Some of the active components are discussed below:

(i) Transistor: Transistor is formed from two words "Transfer" and "Resistor". Thus, it transfers resistance from one part of circuit to other. If at input side the resistance is high then the resistance at the output side will be low. Transistor is a three terminal device which can act as a switch or an amplifier. It can be controlled either by voltage or current thus it is called voltage controlled device or a current controlled device.

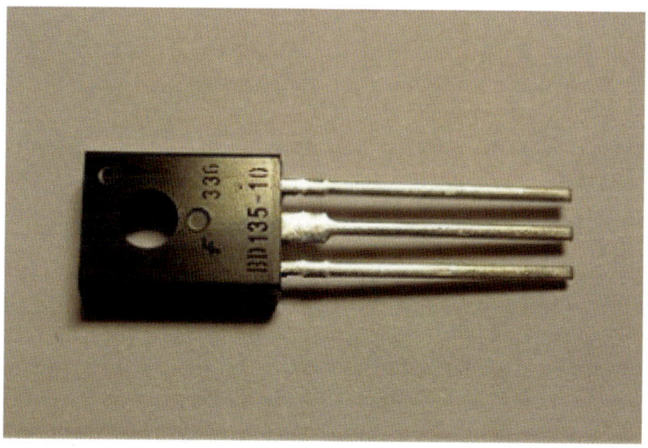

(ii) Batteries: Batteries generate electrical energy by the process of electrochemical reaction which takes place inside it. The chemicals present in the battery undergo transformation to convert chemical energy into electrical energy. It is a vital part of every electronic circuit as it supplied power to the circuit without which the circuit cannot process signals.

(iii) Cathode Ray Tube: It is used in picture tube in television circuit. It is controlled by an electron gun. Electron beam strikes the phosphor screen and it gets illuminated.

Active Elements

The elements that supply energy to the circuit is called active element. Examples of active elements include voltage and current sources, generators, and electronic devices that require power supplies. A transistor is an active circuit element, meaning that it can amplify power of a signal. On the other hand, transformer is not an active element because it does not amplify the power level and power remains same both in primary and secondary sides. Transformer is an example of passive element.

Transistor

Assorted transistors.

A transistor is a semiconductor device that uses a small amount of voltage or electrical current to control a larger change in voltage or current. Because of its fast response and accuracy, it may be used in a wide variety of applications, including amplification, switching, voltage stabilization, signal modulation, and as an oscillator. The transistor is the fundamental building block of both digital and analog circuits—the circuitry that governs the operation of computers, cellular phones, and all other modern electronics. Transistors may be packaged individually or as part of an integrated circuit chip, which may hold thousands of transistors in a very small area.

Modern transistors are divided into two main categories:

- Bipolar junction transistors (BJTs);
- Field effect transistors (FETs).

Application of current in BJTs and voltage in FETs between the input and common terminals increases the conductivity between the common and output terminals, thereby controlling current flow between them.

The term "Transistor" originally referred to the point contact type, but these only saw very limited commercial application, being replaced by the much more practical bipolar junction types in the early 1950s. Ironically both the term "Transistor" itself and the schematic symbol most widely used for it today are the ones that specifically referred to these long-obsolete devices; attempts at introducing more accurate versions have come to nothing.

In analog circuits, transistors are used in amplifiers, (direct current amplifiers, audio amplifiers, radio frequency amplifiers), and linear regulated power supplies. Transistors are also used in digital circuits where they function as electronic switches, but rarely as discrete devices, almost always being incorporated in Monolithic Integrated Circuits. Digital circuits include logic gates, random access memory (RAM), microprocessors, and digital signal processors (DSPs).

Types

Transistors are categorized by:

- Semiconductor material: germanium, silicon, gallium arsenide, silicon carbide;
- Structure: BJT, JFET, IGFET (MOSFET), IGBT, "other types";
- Polarity: NPN, PNP, N-channel, P-channel;
- Maximum power rating: low, medium, high;
- Maximum operating frequency: low, medium, high, radio frequency (RF), microwave (The maximum effective frequency of a transistor is denoted by the term f_T, an abbreviation for "frequency of transition." The frequency of transition is the frequency at which the transistor yields unity gain);

- Application: switch, general purpose, audio, high voltage, super-beta, matched pair;
- Physical packaging: through hole metal, through hole plastic, surface mount, ball grid array.

Thus, a particular transistor may be described as: *silicon, surface mount, BJT, NPN, low power, high frequency switch.*

Bipolar Junction Transistor

The bipolar junction transistor (BJT) was the first type of transistor to be mass-produced. Bipolar transistors are so named because they conduct by using both majority and minority carriers. The three terminals of the BJT are named *emitter, base* and *collector*. Two p-n junctions exist inside a BJT: the *base/emitter junction* and *base/collector junction*. The BJT is commonly described as a current-operated device because the collector/emitter current is controlled by the current flowing between base and emitter terminals. Unlike the FET, the BJT is a low input-impedance device. Because of this exponential relationship the BJT has a higher transconductance than the FET.

Bipolar transistors can be made to conduct by light, since absorption of photons in the base region generates a photocurrent that acts as a base current; the collector current is approximately beta times the photocurrent. Devices designed for this purpose have a transparent window in the package and are called phototransistors.

Bipolar Junction Transistor Biasing

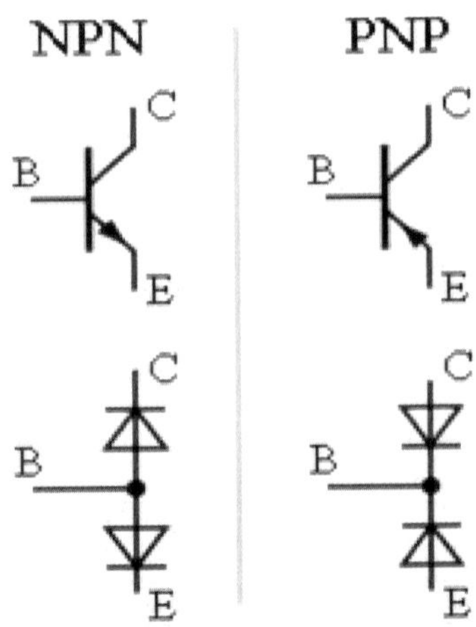

Transistors are the most important semiconductor active devices essential for almost all circuits. They are used as electronic switches, amplifiers etc in circuits. Transistors may be NPN, PNP, FET, JFET etc which have different functions in electronic circuits. For the proper working of the circuit, it is necessary to bias the transistor using resistor networks. Operating point is the point on the output characteristics that shows the Collector-Emitter voltage and the Collector current with no input signal. The Operating point is also known as the Bias point or Q-Point (Quiescent point).

Biasing is referred to provide resistors, capacitors or supply voltage etc to provide proper operating characteristics of the transistors. DC biasing is used to obtain DC collector current at a particular collector voltage. The value of this voltage and current are expressed in terms of the Q-Point. In a transistor amplifier configuration, the IC (max) is the maximum current that can flow through the transistor and VCE (max) is the maximum voltage applied across the device. To work the transistor as an amplifier, a load resistor RC must be connected to the collector. Biasing set the DC operating voltage and current to the correct level so that the AC input signal can be properly amplified by the transistor. The correct biasing point is somewhere between the fully ON or fully OFF states of the transistor. This central point is the Q-Point and if the transistor is properly biased, the Q-point will be the central operating point of the transistor. This helps the output current to increase and decrease as the input signal swings through the complete cycle.

For setting the correct Q-Point of the transistor, a collector resistor is used to set the collector current to a constant and steady value without any signal in its base. This steady DC operating point is set by the value of the supply voltage and the value of the base biasing resistor. Base bias resistors are used in all the three transistor configurations like common base, common collector and Common emitter configurations.

Modes of Biasing

Following are the different modes of transistor base biasing:

Current Biasing

As shown in the figure, two resistors RC and RB are used to set the base bias. These resistors establish the initial operating region of the transistor with a fixed current bias.

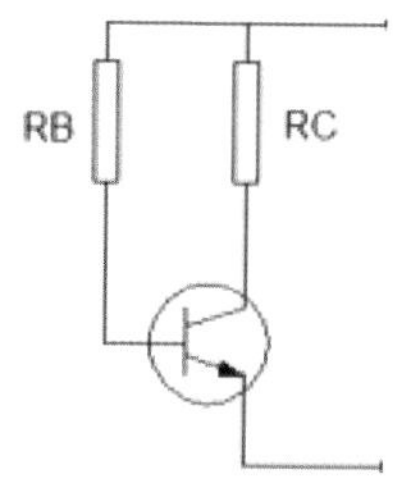

Current biasing

The transistor forward biases with a positive base bias voltage through RB. The forward base-Emitter voltage drop is 0.7 volts. Therefore the current through RB is $I_B = (V_{cc} - V_{BE}) / I_B$.

Feedback Biasing

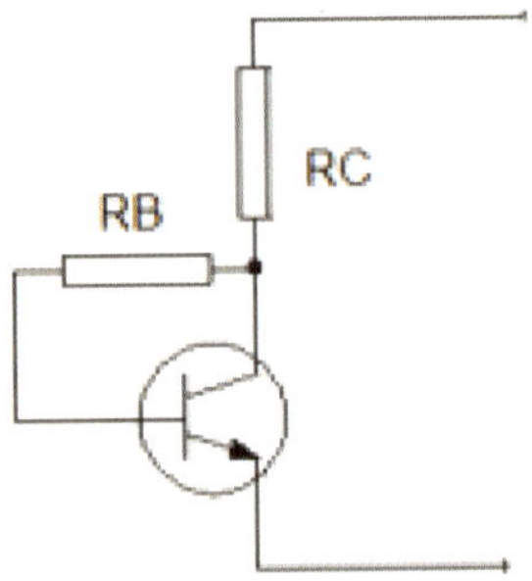

Feedback biasing

Figure shows the transistor biasing by the use of a feedback resistor. The base bias is obtained from the collector voltage. The collector feedback ensures that the transistor is always biased in the active region. When the collector current increases, the voltage at the collector drops. This reduces the base drive which in turn reduces the collector current. This feedback configuration is ideal for transistor amplifier designs.

Double Feedback Biasing

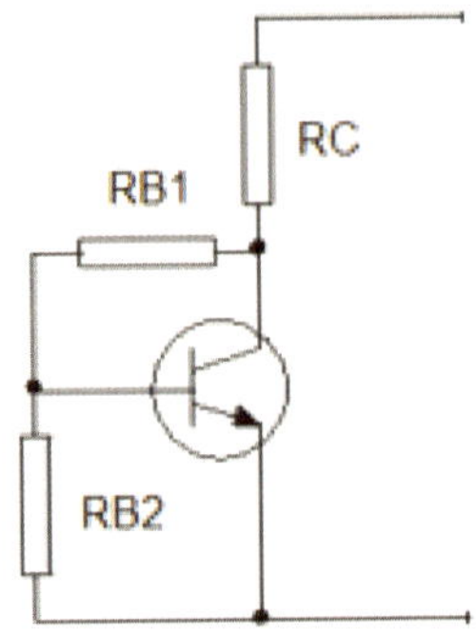

Double Feedback biasing

Figure shows how the biasing is achieved using double feedback resistors.

By using two resistors RB1 and RB2 increases the stability with respect to the variations in Beta by increasing the current flow through the base bias resistors. In this configuration, the current in RB1 is equal to 10 % of the collector current.

Voltage Dividing Biasing

Figure shows the Voltage divider biasing in which two resistors RB1 and RB2 are connected to the base of the transistor forming a voltage divider network. The transistor gets biases by the voltage drop across RB2. This kind of biasing configuration is used widely in amplifier circuits.

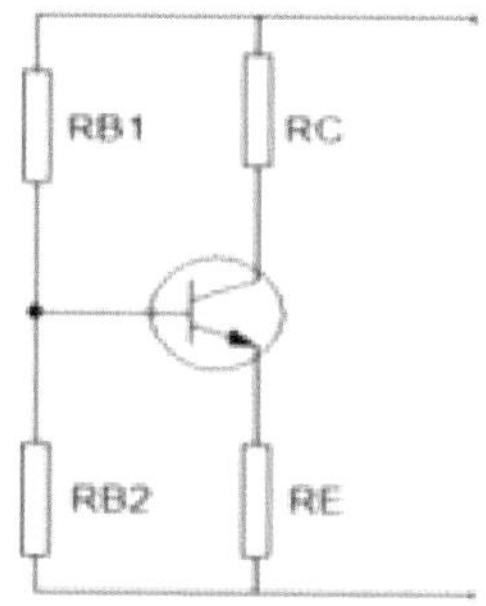

Voltage divider biasing

Double Base Biasing

Figure shows a double feedback for stabilization. It uses both Emitter and Collector base feedback to improve the stabilization through controlling the collector current. Resistor values should be selected so as to set the voltage drop across the Emitter resistor 10% of the supply voltage and the current through RB1, 10% of the collector current.

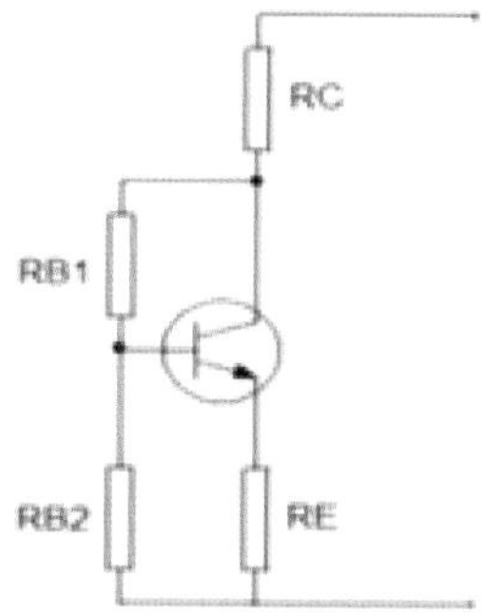

Double base biaisng

Field-effect Transistor

The field-effect transistor (FET), sometimes called a *unipolar transistor,* uses either electrons (N-channel FET) or holes (P-channel FET) for conduction. The four terminals of the FET are named *source, gate, drain,* and *body (substrate).* On most FETs the body is connected to the source inside the package and this will be assumed for the following description.

A voltage applied between the gate and source (body) controls the current flowing between the drain and source. As the gate/source voltage (Vgs) is increased the drain/ source current (Ids) increases parabolically. In FETs the drain/source current flows through a conducting channel near the *gate*. This channel connects the *drain* region to the *source* region. The channel conductivity is varied by the electric field generated by the voltage applied between the gate/source terminals. In this way the current flowing between the drain and source is controlled.

FETs are divided into two families: junction FET (JFET) and insulated gate FET (IGFET). The IGFET is more commonly known as metal–oxide–semiconductor FET (MOSFET), from their original construction as a layer of metal (the gate), a layer of oxide (the insulation), and a layer of semiconductor. Unlike IGFETs, the JFET gate forms a PN diode with the channel which lies between the source and drain. Functionally, this makes the N-channel JFET the solid state equivalent of the vacuum tube triode which, similarly, forms a diode between its grid and cathode. Also, both devices operate in the *depletion mode,* they both have a high input impedance, and they both conduct current under the control of an input voltage.

MESFETs are JFETs, in which the reverse biased PN junction is replaced by a semiconductor-metal Schottky-junction. These, and the HEMFETs (high electron mobility FETs), in which a two-dimensional electron gas with very high carrier mobility is used for charge transport, are especially suitable for use at very high frequencies (microwave frequencies; several GHz).

Unlike bipolar transistors, FETs do not inherently amplify a photocurrent. Nevertheless, there are ways to use them, especially JFETs, as light-sensitive devices, by exploiting the photocurrents in channel–gate or channel–body junctions.

FETs are further divided into depletion-mode and enhancement-mode types, depending on whether the channel is turned on or off with zero gate-to-source voltage. For enhancement mode, the channel is off at zero bias, and a gate potential can "enhance" the conduction. For depletion mode, the channel is on at zero bias, and a gate potential (of the opposite polarity) can "deplete" the channel, reducing conduction. For either mode, a more positive gate voltage corresponds to a higher current for N-channel devices and a lower current for P-channel devices. Nearly all JFETs are depletion-mode as the diode junctions would forward bias and conduct if they were enhancement mode devices; most IGFETs are enhancement-mode types.

Other Transistor Types

- Heterojunction Bipolar Transistor (HBT) is an improvement of the bipolar junction transistor (BJT) that can handle signals of very high frequencies up to several hundred GHz. It is common in modern ultrafast circuits, mostly radio-frequency (RF) systems.
- Unijunction transistors can be used as simple pulse generators. They comprise a main body of either P-type or N-type semiconductor with ohmic contacts at each end (terminals *Base1* and *Base2*). A junction with the opposite semiconductor type is formed at a point along the length of the body for the third terminal *(Emitter).*
- Dual gate FETs have a single channel with two gates in cascode; a configuration that is optimized for high-frequency amplifiers, mixers, and oscillators.
- Transistor arrays are used for general purpose applications, function generation,

and low-level, low-noise amplifiers. They include two or more transistors on a common substrate to ensure close parameter matching and thermal tracking, characteristics that are especially important for long tailed pair amplifiers.

- Darlington transistors comprise a medium power BJT connected to a power BJT. This provides a high current gain equal to the product of the current gains of the two transistors. Power diodes are often connected between certain terminals depending on specific use.
- Insulated Gate Bipolar Transistor (IGBT transistor) use a medium power IGFET, similarly connected to a power BJT, to give a high input impedance. Power diodes are often connected between certain terminals depending on specific use. IGBTs are particularly suitable for heavy-duty industrial applications.
- Single-electron transistors (SET) consist of a gate island between two tunnelling junctions. The tunnelling current is controlled by a voltage applied to the gate through a capacitor.
- Nanofluidic Transistor Control the movement of ions through sub-microscopic, water-filled channels. Nanofluidic transistor, the basis of future chemical processors.
- Trigate transistors (Prototype by Intel, also known as three dimensional transistors) use a single gate that is stacked on top of two vertical gates allowing for essentially three times the surface area for electrons to travel.
- Avalanche transistors have the ability to switch very high currents with less than a nanosecond rise and fall times (transition times).
- Ballistic transistor, Electrons bounce their way through maze.
- Spin transistors are magnetically sensitive devices.
- Thin film transistors are used in LCD display.
- Floating gate transistors are used for non-volatile storage.
- Photo transistors react to light.
- Inverted-T field effect transistor, part of the device extends vertically from the horizontal plane in an inverted T shape, hence the name.
- Ion sensitive field effect transistors measure ion concentrations in solution.
- FinFET The source/drain region forms fins on the silicon surface.
- FREDFET Fast-Reverse Epitaxal Diode Field-Effect Transistor.
- EOSFET Electrolyte-Oxide-Semiconductor Field Effect Transistor (Neurochip).

Semiconductor Materials

The first BJTs were made from germanium (Ge), and some high-power types still are. Silicon (Si) types currently predominate but certain advanced microwave and high performance versions now employ the compound semiconductor material gallium arsenide (GaAs) and the semiconductor alloy silicon germanium (SiGe). Single-element semiconductor materials (Ge or Si) are described as "elemental."

Packaging

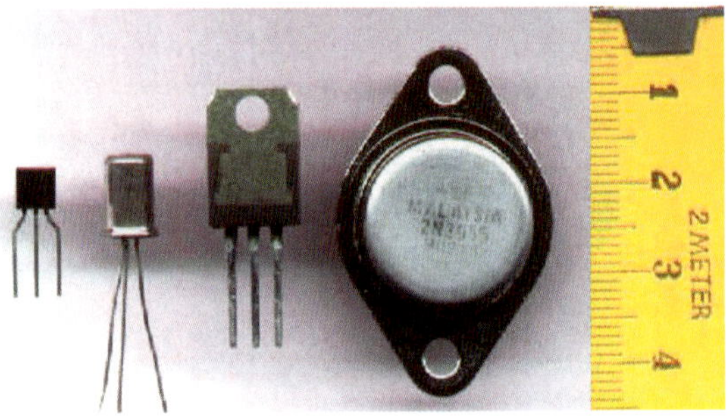

Through-hole transistors (tape measure marked in centimetres)

Transistors come in many different packages (chip carriers). The two main categories are *through-hole* (or *leaded*), and *surface-mount,* also known as *surface mount device* (Surface-mount technology, SMD). The "ball grid array" (BGA) is the latest surface mount package (currently only for large transistor arrays). It has solder "balls" on the underside in place of leads. Because they are smaller and have shorter interconnections, SMDs have better high frequency characteristics but lower power rating.

Transistor packages are made of glass, metal, ceramic or plastic. The package often dictates the power rating and frequency characteristics. Power transistors have large packages that can be clamped to heat sinks for enhanced cooling. Additionally, most power transistors have the collector or drain physically connected to the metal can/metal plate. At the other extreme, some surface-mount "microwave" transistors are as small as grains of sand.

Often a given transistor type is available in different packages. Transistor packages are mainly standardized, but the assignment of a transistor's functions to the terminals is not: different transistor types can assign different functions to the package's terminals.

Transistor Construction

Transistors rely on semiconductors to work their magic. A semiconductor is a material that's not quite a pure conductor (like copper wire) but also not an insulator (like air). The conductivity of a semiconductor – how easily it allows electrons to flow – depends on variables like temperature or the presence of more or less electrons. Let's look briefly under the hood of a transistor. Don't worry, we won't dig too deeply into quantum physics.

A Transistor as Two Diodes

Transistors are kind of like an extension of another semiconductor component: diodes. In a way transistors are just two diodes with their cathodes (or anodes) tied together:

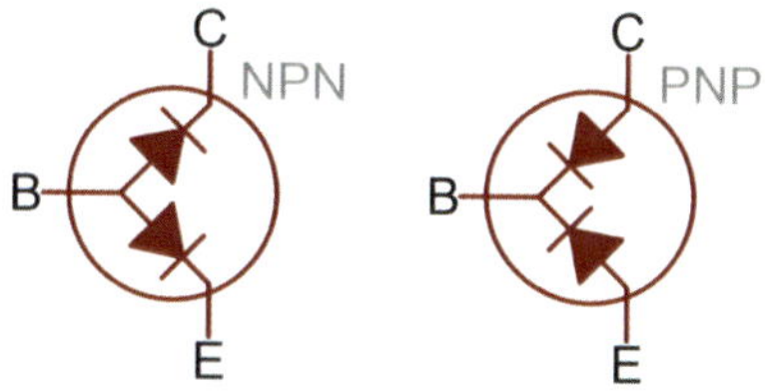

The diode connecting base to emitter is the important one here; it matches the direction of the arrow on the schematic symbol, and shows you which way current is intended to flow through the transistor.

The diode representation is a good place to start, but it's far from accurate. Don't base your understanding of a transistor's operation on that model (and definitely don't try to replicate it on a breadboard, it won't work). There's a whole lot of weird quantum physics level stuff controlling the interactions between the three terminals.

(This model *is* useful if you need to test a transistor. Using the diode (or resistance) test function on a multimeter, you can measure across the BE and BC terminals to check for the presence of those "diodes").

Transistor Structure and Operation

Transistors are built by stacking three different layers of semiconductor material together. Some of those layers have extra electrons added to them (a process called "doping"), and others have electrons removed (doped with "holes" – the absence of electrons). A semiconductor material with *extra* electrons is called an n-type (*n* for negative because electrons have a negative charge) and a material with electrons removed is called a p-type (for positive). Transistors are created by either stacking an *n* on top of a *p* on top of an *n*, or *p* over *n* over *p*.

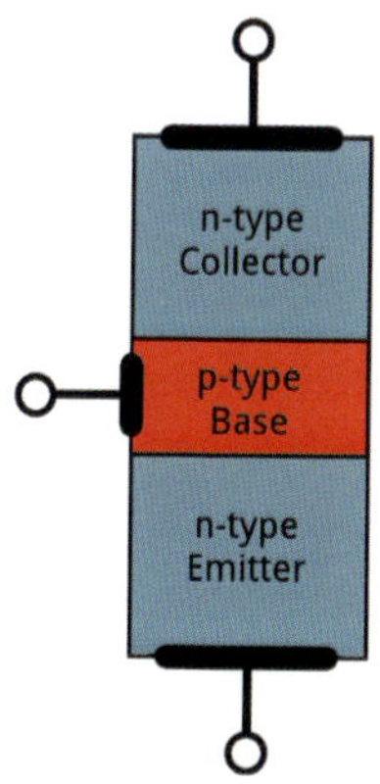

Simplified diagram of the structure of an NPN

With some hand waving, we can say electrons can easily flow from *n* regions to *p* regions, as long as they have a little force (voltage) to push them. But flowing from a *p* region to an *n* region is really hard (requires a *lot* of voltage). But the special thing about a transistor – the part that makes our two-diode model obsolete – is the fact that electrons *can* easily flow from the p-type base to the n-type collector as long as the base-emitter junction is forward biased (meaning the base is at a higher voltage than the emitter).

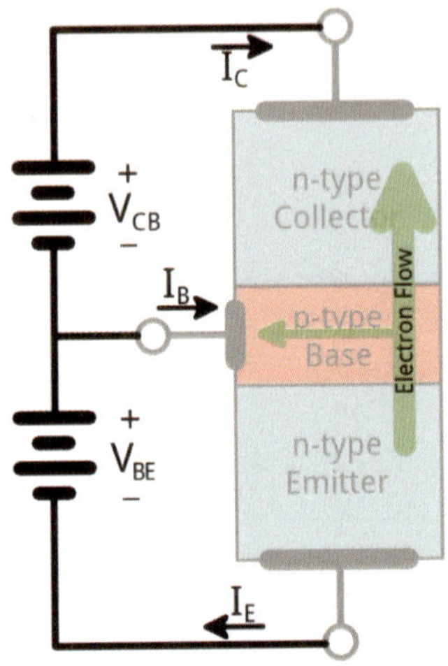

The NPN transistor is designed to pass electrons from the emitter to the collector (so conventional current flows from collector to emitter). The emitter "emits" electrons into the base, which controls the number of electrons the emitter emits. Most of the electrons emitted are "collected" by the collector, which sends them along to the next part of the circuit.

A PNP works in a same but opposite fashion. The base still controls current flow, but that current flows in the opposite direction – from emitter to collector. Instead of electrons, the emitter emits "holes" (a conceptual absence of electrons) which are collected by the collector.

The transistor is kind of like an electron valve. The base pin is like a handle you might adjust to allow more or less electrons to flow from emitter to collector.

Usage

In the early days of transistor circuit design, the bipolar junction transistor (or BJT) was the most commonly used transistor. Even after MOSFETs became available, the BJT remained the transistor of choice for digital and analog circuits because of their ease of manufacture and speed. However, the MOSFET has several desirable properties for digital circuits, and major advances in digital circuits have pushed MOSFET design to state-of-the-art. MOSFETs are now commonly used for both analog and digital functions.

Switches

Transistors are commonly used as electronic switches, for both high-power applications including switched-mode power supplies and low-power applications such as logic gates.

Amplifiers

From mobile phones to televisions, vast numbers of products include amplifiers for sound reproduction, radio transmission, and signal processing. The first discrete transistor audio amplifiers barely supplied a few hundred milliwatts, but power and audio fidelity gradually increased as better transistors became available and amplifier architecture evolved.

Transistors are commonly used in modern musical instrument amplifiers, where circuits up to a few hundred watts are common and relatively cheap. Transistors have largely replaced valves in instrument amplifiers. Some musical instrument amplifier manufacturers mix transistors and vacuum tubes in the same circuit, to utilize the inherent benefits of both devices.

Computers

The "first generation" of electronic computers used vacuum tubes, which generated large amounts of heat and were bulky, and unreliable. The development of the transistor was key to computer miniaturization and reliability. The "second generation" of computers, through the late 1950s and 1960s, featured boards filled with individual transistors and magnetic memory cores. Subsequently, transistors, other components, and their necessary wiring were integrated into a single, mass-manufactured component: the integrated circuit. Transistors incorporated into integrated circuits have replaced most discrete transistors in modern digital computers.

Importance

The transistor is considered by many to be one of the greatest inventions in modern history, ranking in importance with the printing press, car, and telephone. It is the key active component in practically all modern electronics. Its importance in today's society rests on its ability to be mass produced using a highly automated process (fabrication) that achieves vanishingly low per-transistor costs.

Although millions of individual (known as *discrete*) transistors are still used, the vast majority of transistors are fabricated into integrated circuits (often abbreviated as *IC* and also called *microchips* or simply *chips*) along with diodes, resistors, capacitors and other electronic components to produce complete electronic circuits. A logic gate comprises of about twenty transistors whereas an advanced microprocessor, as of 2006, can use as many as 1.7 billion transistors (MOSFETs).

The transistor's low cost, flexibility and reliability have made it a universal device for non-mechanical tasks, such as digital computing. Transistorized circuits have replaced electromechanical devices for the control of appliances and machinery as well. It is often less expensive and more effective to use a standard microcontroller and write a computer program to carry out a control function than to design an equivalent mechanical control function.

Because of the low cost of transistors and hence digital computers, there is a trend to digitize information. With digital computers offering the ability to quickly find, sort and process digital information, more and more effort has been put into making information digital. As a result, today, much media data is delivered in digital form, finally being converted and presented in analog form by computers. Areas influenced by the Digital Revolution include television, radio, and newspapers.

Advantages of Transistors Over Vacuum Tubes

Before the development of transistors, vacuum tubes (or in the UK thermionic valves or just valves) were the main active components in electronic equipment. The key advantages that have allowed transistors to replace their vacuum tube predecessors in most applications are:

- Smaller size (despite continuing miniaturization of vacuum tubes);
- Highly automated manufacture;
- Lower cost (in volume production);
- Lower possible operating voltages (but vacuum tubes can operate at higher voltages);
- No warm-up period (most vacuum tubes need 10 to 60 seconds to function correctly);
- Lower power dissipation (no heater power, very low saturation voltage);
- Higher reliability and greater physical ruggedness (although vacuum tubes are electrically more rugged. Also the vacuum tube is much more resistant to nuclear electromagnetic pulses (NEMP) and electrostatic discharge (ESD));
- Much longer life (vacuum tube cathodes are eventually exhausted and the vacuum can become contaminated);
- Complementary devices available (allowing circuits with complementary-symmetry: vacuum tubes with a polarity equivalent to PNP BJTs or P type FETs are not available);
- Ability to control large currents (power transistors are available to control hundreds of amperes, vacuum tubes to control even one ampere are large and costly);

- Much less microphonic (vibration can modulate vacuum tube characteristics, though this may contribute to the sound of guitar amplifiers).

Batteries

Batteries are a collection of one or more cells whose chemical reactions create a flow of electrons in a circuit. All batteries are made up of three basic components: an anode (the '-' side), a cathode (the '+' side), and some kind of electrolyte (a substance that chemically reacts with the anode and cathode).

When the anode and cathode of a battery is connected to a circuit, a chemical reaction takes place between the anode and the electrolyte. This reaction causes electrons to flow through the circuit and back into the cathode where another chemical reaction takes place. When the material in the cathode or anode is consumed or no longer able to be used in the reaction, the battery is unable to produce electricity. At that point, your battery is "dead."

Batteries that must be thrown away after use are known as primary batteries. Batteries that can be recharged are called secondary batteries.

Lithium polymer batteries, for example, can be recharged

Without batteries, your quadcopter would have to be tethered to the wall, you would have to hand crank your car, and your Xbox controller would have to plugged in all the time (like in the good old days). Batteries offer a way to store electrical potential energy in a portable container.

Batteries come in a variety of shapes, sizes, and chemistries.

The invention of the modern battery is often attributed to Alessandro Volta. It actually started with a surprising accident involving the dissection of a frog.

Diode

A diode is an electrical device allowing current to move through it in one direction with far greater ease than in the other. The most common kind of diode in modern circuit design is the semiconductor diode, although other diode technologies exist. Semiconductor diodes are symbolized in schematic diagrams such as figure below. The term "diode" is customarily reserved for small signal devices, I ≤ 1 A. The term rectifier is used for power devices, I > 1 A.

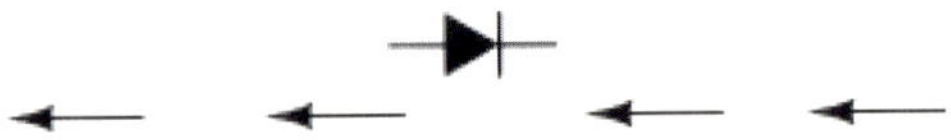

Semiconductor diode schematic symbol: Arrows indicate the direction of electron current flow.

When placed in a simple battery-lamp circuit, the diode will either allow or prevent current through the lamp, depending on the polarity of the applied voltage.

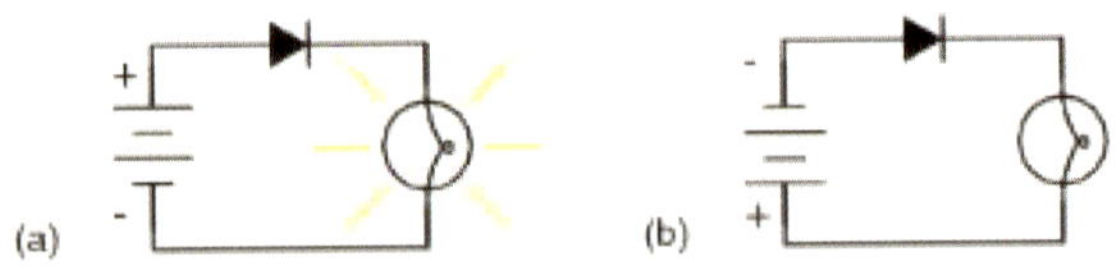

Diode operation: (a) Current flow is permitted; the diode is forward biased.
(b) Current flow is prohibited; the diode is reversed biased.

When the polarity of the battery is such that electrons are allowed to flow through the diode, the diode is said to be *forward-biased*. Conversely, when the battery is "backward" and the diode blocks current, the diode is said to be *reverse-biased*. A diode may be thought of as like a switch: "closed" when forward-biased and "open" when reverse-biased.

Oddly enough, the direction of the diode symbol's "arrowhead" points *against* the direction of electron flow. This is because the diode symbol was invented by engineers, who predominantly use *conventional flow* notation in their schematics, showing current as a flow of charge from the positive (+) side of the voltage source to the negative (-). This convention holds true for all semiconductor symbols possessing "arrowheads:" the arrow points in the permitted direction of conventional flow, and against the permitted direction of electron flow.

Diode behaviour is analogous to the behaviour of a hydraulic device called a *check valve*. A check valve allows fluid flow through it in only one direction as in figure below.

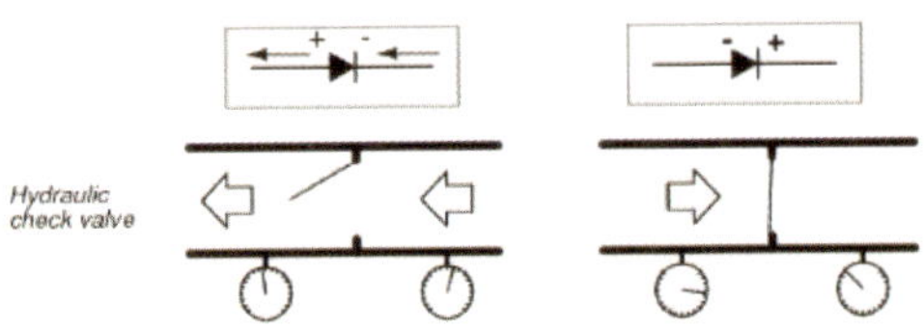

Hydraulic check valve analogy: (a) Electron current flow permitted. (b) Current flow prohibited.

Check valves are essentially pressure-operated devices: they open and allow flow if the pressure across them is of the correct "polarity" to open the gate (in the analogy shown, greater fluid pressure on the right than on the left). If the pressure is of the opposite "polarity," the pressure difference across the check valve will close and hold the gate so that no flow occurs.

Like check valves, diodes are essentially "pressure-" operated (voltage-operated) devices. The essential difference between forward-bias and reverse-bias is the polarity of the voltage dropped across the diode. Let's take a closer look at the simple battery-diode-lamp circuit shown earlier, this time investigating voltage drops across the various components in figure below.

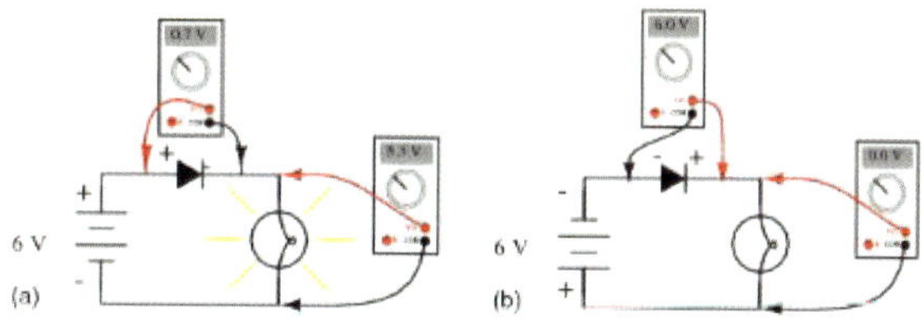

Diode circuit voltage measurements: (a) Forward biased. (b) Reverse biased.

A forward-biased diode conducts current and drops a small voltage across it, leaving most of the battery voltage dropped across the lamp. If the battery's polarity is reversed, the diode becomes reverse-biased, and drops *all* of the battery's voltage leaving none for the lamp. If we consider the diode to be a self-actuating switch (closed in the forward-bias mode and open in the reverse-bias mode), this behaviour makes sense. The most substantial difference is that the diode drops a lot more voltage when conducting than the average mechanical switch (0.7 volts versus tens of millivolts).

This forward-bias voltage drop exhibited by the diode is due to the action of the depletion region formed by the P-N junction under the influence of an applied voltage. If no voltage applied is across a semiconductor diode, a thin depletion region exists around the region of the P-N junction, preventing current flow. (figure below (a)) The depletion region is almost devoid of available charge carriers, and acts as an insulator:

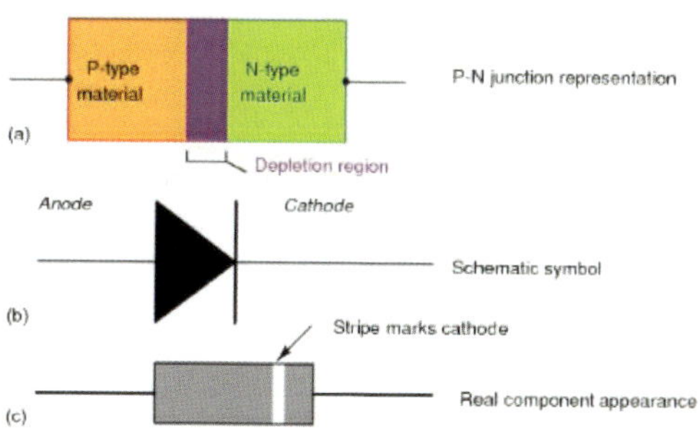

Diode representations: PN-junction model, schematic symbol, physical part.

The schematic symbol of the diode is shown in figure above (b) such that the anode (pointing end) corresponds to the P-type semiconductor at (a). The cathode bar, non-pointing end, at (b) corresponds to the N-type material at (a). Also note that the cathode stripe on the physical part (c) corresponds to the cathode on the symbol.

If a reverse-biasing voltage is applied across the P-N junction, this depletion region expands, further resisting any current through it. (figure below)

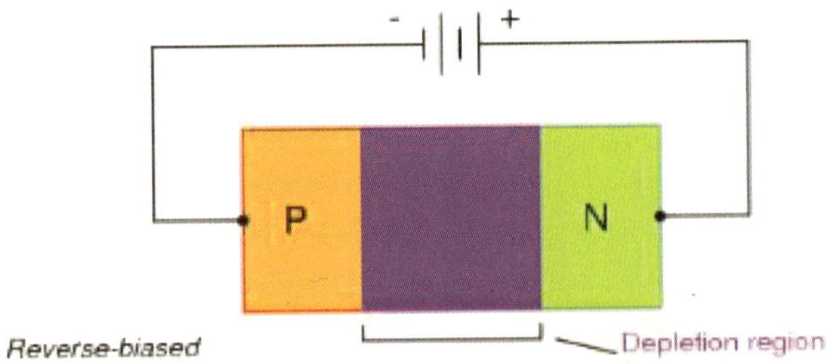

Depletion region expands with reverse bias.

Conversely, if a forward-biasing voltage is applied across the P-N junction, the depletion region collapses becoming thinner. The diode becomes less resistive to current through it. In order for a sustained current to go through the diode; though, the depletion region must be fully collapsed by the applied voltage. This takes a certain minimum voltage to accomplish, called the *forward voltage* as illustrated in figure below.

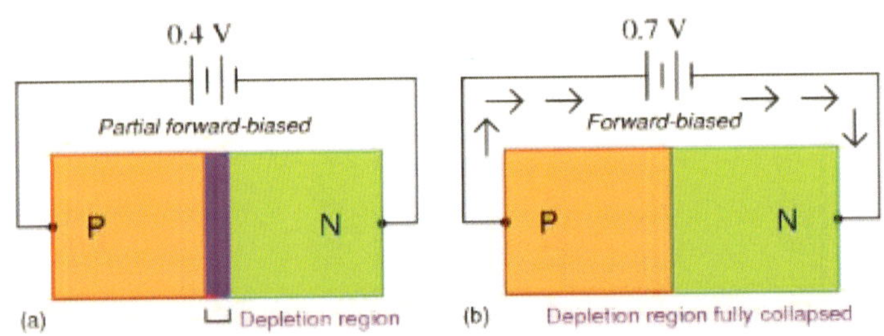

Inceasing forward bias from (a) to (b) decreases depletion region thickness.

For silicon diodes, the typical forward voltage is 0.7 volts, nominal. For germanium diodes, the forward voltage is only 0.3 volts. The chemical constituency of the P-N junction comprising the diode accounts for its nominal forward voltage figure, which is why silicon and germanium diodes have such different forward voltages. Forward voltage drop remains approximately constant for a wide range of diode currents, meaning that diode voltage drop is not like that of a resistor or even a normal (closed) switch. For most simplified circuit analysis, the voltage drop across a conducting diode may be considered constant at the nominal figure and not related to the amount of current.

Actually, forward voltage drop is more complex. An equation describes the exact current through a diode, given the voltage dropped across the junction, the temperature of the junction, and several physical constants. It is commonly known as the *diode equation*:

$$I_D = I_S\left(e^{qV_D/NkT} - 1\right)$$

Where,

$$\begin{aligned}
I_D &= \text{Diode current in amps}\\
I_S &= \text{Saturation current in amps (typically } 1\times10^{-12}\text{ amp)}\\
e &= \text{Euler's constant } (\sim 2.718281828)\\
q &= \text{Change of electron } (1.6\times10^{-19}\text{ coulombs})\\
V_D &= \text{Voltage applied across diode in volts}\\
N &= \text{"Nonideality" or "emission" coefficient (typically between 1 and 2)}\\
k &= \text{Boltzmann's constant } (1.38\times10^{-23})\\
T &= \text{Junction temperature in Kelvins}
\end{aligned}$$

The term kT/q describes the voltage produced within thc P-N junction due to the action of temperature, and is called the *thermal voltage*, or V_t of the junction. At room temperature, this is about 26 millivolts. Knowing this, and assuming a "nonideality" coefficient of 1, we may simplify the diode equation and re-write it as such:

$$I_D = I_S\left(e^{V_D/0.026} - 1\right)$$

Where,

$$\begin{aligned}
I_D &= \text{Diode current in amps}\\
I_S &= \text{Saturation current in amps (typically } 1\times10^{-12}\text{ amp)}\\
e &= \text{Euler's constant } (\sim 2.718281828)\\
V_D &= \text{Voltage applied across diode in volts}
\end{aligned}$$

You need not be familiar with the "diode equation" to analyze simple diode circuits. Just understand that the voltage dropped across a current-conducting diode *does* change with the amount of current going through it, but that this change is fairly small over a wide range of currents. This is why many textbooks simply say the voltage drop across a conducting, semiconductor diode remains constant at 0.7 volts for silicon and 0.3 volts for germanium. However, some circuits intentionally make use of the P-N junction's inherent exponential current/voltage relationship and thus can only be understood in the context of this equation. Also, since temperature is a factor in the diode equation, a forward-biased P-N junction may also be used as a temperature-sensing device, and thus can only be understood if one has a conceptual grasp on this mathematical relationship.

A reverse-biased diode prevents current from going through it, due to the expanded depletion region. In actuality, a very small amount of current can and does go through a reverse-biased diode, called the *leakage current*, but it can be ignored for most purposes.

The ability of a diode to withstand reverse-bias voltages is limited, as it is for any insulator. If the applied reverse-bias voltage becomes too great, the diode will experience a condition known as *breakdown* (figure below), which is usually destructive. A diode's maximum reverse-bias voltage rating is known as the *Peak Inverse Voltage*, or *PIV*, and may be obtained from the manufacturer. Like forward voltage, the PIV rating of a diode varies with temperature, except that PIV *increases* with increased temperature and *decreases* as the diode becomes cooler—exactly opposite that of forward voltage.

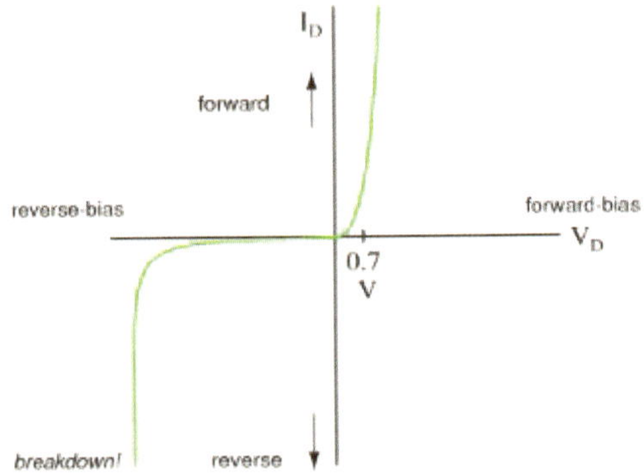

Diode curve: showing knee at 0.7 V forward bias for Si, and reverse breakdown.

Typically, the PIV rating of a generic "rectifier" diode is at least 50 volts at room temperature. Diodes with PIV ratings in the many thousands of volts are available for modest prices.

Passive Elements

The element which receives energy (or absorbs energy) and then either converts it into heat (R) or stored it in an electric (C) or magnetic (L) field is called passive element.

Resistor

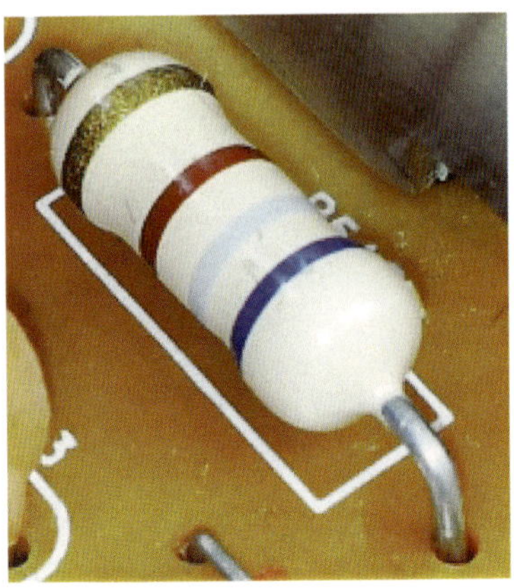

The resistor is a passive electrical component to create resistance in the flow of electric current. In almost all electrical networks and electronic circuits they can be found. The resistance is measured in ohms. An ohm is the resistance that occurs when a current of one ampere passes through a resistor with a one volt drop across its terminals. The current is proportional to the voltage across the terminal ends. This ratio is represented by Ohm's law:

$$R = \frac{V}{I}$$

Resistors are used for many purposes. A few examples include delimit electric current, voltage division, heat generation, matching and loading circuits, control gain, and fix time constants. They are commercially available with resistance values over a range of more than nine orders of magnitude. They can be used to as electric brakes to dissipate kinetic energy from trains, or be smaller than a square millimeter for electronics.

Resistor Definition and Symbol

A resistor is a passive electrical component with the primary function to limit the flow of electric current.

The international IEC symbol is a rectangular shape. In the USA the ANSI standard is very common, this is a zigzag line.

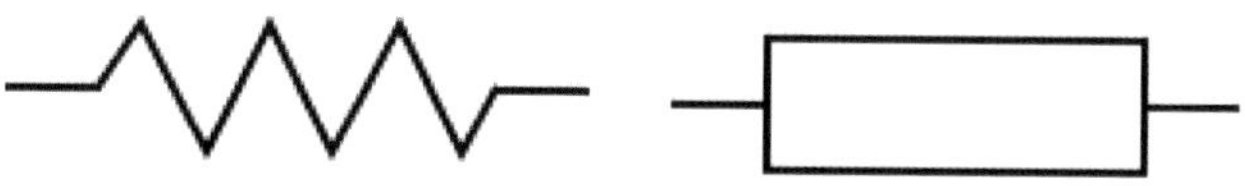

Fixed resistor symbol
ANSI standard

Fixed resistor symbol

Ohm's Law

The resistor's current I in amps (A) is equal to the resistor's voltage V in volts (V) divided by the resistance R in ohms (Ω):

$$I = \frac{V}{R}$$

The resistor's power consumption P in watts (W) is equal to the resistor's current I in amps (A), times the resistor's voltage V in volts (V):

$$P = I \times V$$

The resistor's power consumption P in watts (W) is equal to the square value of the resistor's current I in amps (A), times the resistor's resistance R in ohms (Ω):

$$P = I^2 \times R$$

The resistor's power consumption P in watts (W) is equal to the square value of the resistor's voltage V in volts (V) divided by the resistor's resistance R in ohms (Ω):

$$P = V^2 / R$$

Resistor Types

There are many different types of resistor available for use within electronic circuits. These different resistor types have somewhat different properties dependent upon their construction and manufacture. This makes the different types of resistor suitable for different applications.

Over the years the resistor types used in mass electronics production have changed. Years ago, all the resistors used had leads and were relatively large, and by today's standards they offered a low level of performance. Today, the resistor types used are much smaller and offer much higher levels of performance.

Fixed and Variable Resistor Types

The first major categories into which the different types of resistor can be fitted is into whether they are fixed or variable:

- *Fixed resistors:* Fixed resistors are by far the most widely used type of resistor. They are used in electronics circuits to set the right conditions in a circuit. Their values are determined during the design phase of the circuit, and they should never need to be changed to "adjust" the circuit. There are many different types of resistor which can be used in different circumstances and these different types of resistor are described in further detail below.
- *Variable resistors:* These resistors consist of a fixed resistor element and a slider which taps onto the main resistor element. This gives three connections to the component: two connected to the fixed element, and the third is the slider. In this way the component acts as a variable potential divider if all three connections are used. It is possible to connect to the slider and one end to provide a resistor with variable resistance.

Fixed Resistor Types

There are a number of different types of fixed resistor:

- Carbon composition: These types were once very common, but are now seldom used. They are formed by mixing carbon granules with a binder which was then made into a small rod. This type of resistor was large by today's standards and suffered from a large negative temperature coefficient. The resistors also suffered from a large and erratic irreversible changes in resistance as a result of heat or age. In addition to this the granular nature of the carbon and binder lead to high levels of noise being generated when current flowed.

- Carbon film: This resistor type is formed by "cracking" a hydrocarbon onto a ceramic former. The resulting deposited film had its resistance set by cutting a helix into the film. This made these resistors highly inductive and of little use for many RF applications. They exhibited a temperature coefficient of between -100 and -900 ppm/°Celcius. The carbon film is protected either by a conformal epoxy coating or a ceramic tube. *Metal oxide film:* This type of resistor is now one of the most widely used form of resistor along with the metal film type. Rather than using a carbon film, this resistor type uses a metal oxide film deposited on a ceramic rod. Metals oxide such as tin oxide are deposited onto the ceramic rod. The resistance of the component is adjusted in two ways. First the thickness of the deposited layer is controlled during the initial manufacturing stages. Then it can be more accurately adjusted by cutting a helical grove in the film. Again the film is protected using a conformal epoxy coating. This type of resistor has a temperature coefficient of around ±15 parts per million/°K, giving it a far superior performance to that of any carbon based resistor. Additionally this type of resistor can be supplied to a much closer tolerance, ±5%, ±2% being standard, and with ±1% versions available. They also exhibit a much lower noise level than carbon types of resistor.
- Metal film: The metal film resistors is very similar to the metal oxide film resistor in terms of visual appearance and performance. Instead of using a metal oxide film, this type of resistor uses a metal film. Metals such as nickel alloy may be used.
- Wire wound: This resistor type is generally reserved for high power applications. These resistors are made by winding wire with a higher than normal resistance (resistance wire) on a former. The more expensive varieties are wound on a ceramic former and they may be covered by a vitreous or silicone enamel. This resistor type is suited to high powers and exhibits a high level of reliability at high powers along with a comparatively low level of temperature coefficient, although this will depend on a number of factors including the former, wire used, etc.
- Thin film: Thin film technology is used for most of the surface mount types of resistor. As these are used in their billions these days, this makes this form of resistor technology one of the most widely used.

Leaded and Non-leaded Resistor Types

One of the key differentiators for resistors, and many other forms of component these days is the way in which they are connected. As a result of the mass production techniques sued and the widespread use of printed circuit boards, the form of connection used for components, especially those to be incorporated into mass produced items changed.

The two main forms of resistor type according to their connection method are:

- Leaded resistors: This type of resistor has been used since the very first electronic components have been in use. Typically components were connected to terminal posts of one form or another and leads from the resistor element were needed. As time progressed, printed circuit boards were used, and the leads were inserted through holes in the boards and typically soldered on the reverse side where the tracks were to be found.

Typical leaded carbon resistor

- Surface mount resistors: These resistor types have been used increasingly since the introduction of surface mount technology. Typically this type of resistor is manufactured using thin film technology. A full range of values can be obtained.

Typical SMD resistors on a PCB

Identifying Resistors

Most axial resistors use a pattern of colored stripes to indicate resistance. Surface-mount resistors are marked numerically. Cases are usually tan, brown, blue, or green, though other colors are occasionally found, such as dark red or dark gray.

One can also use a multimeter or ohmmeter to test the resistance values of a resistor.

Four-band Axial Resistors

Four-band identification is the most commonly used color coding scheme on all resistors. It consists of four colored bands that are painted around the body of the resistor.

The first two bands encode the first two significant digits of the resistance value, the third is a power-of-ten multiplier, and the fourth is the tolerance, or acceptable error, of the value.

For example, green-blue-yellow-red is $56\times10^4\ \Omega = 560\ k\Omega \pm 2$ percent).

Each color corresponds to a certain number, shown in the chart below. The tolerance for a 4-band resistor will be 1 percent, 5 percent, or 10 percent.

Color	1st band	2nd band	3rd band (multiplier)	4th band (tolerance)	Temp. Coefficient
Black	0	0	$\times10^0$		
Brown	1	1	$\times10^1$	±1% (F)	100 ppm
Red	2	2	$\times10^2$	±2% (G)	50 ppm
Orange	3	3	$\times10^3$		15 ppm
Yellow	4	4	$\times10^4$		25 ppm
Green	5	5	$\times10^5$	±0.5% (D)	
Blue	6	6	$\times10^6$	±0.25% (C)	
Violet	7	7	$\times10^7$	±0.1% (B)	
Gray	8	8	$\times10^8$	±0.05% (A)	
White	9	9	$\times10^9$		
Gold			$\times10^{-1}$	±5% (J)	
Silver			$\times10^{-2}$	±10% (K)	
None				±20% (M)	

Preferred Values

Resistors are manufactured in values from a few milliohms to about a gigaohm; only a limited range of values from the IEC 60063 preferred number series are commonly available. These series are called E6, E12, E24, E96, and E192. The number tells how many standardized values exist in each decade (for example, between 10 and 100, or between 100 and 1000). So resistors conforming to the E12 series can have 12 distinct values between 10 and 100, whereas those confirming to the E24 series would have 24 distinct values.

In practice, the discrete component sold as a "resistor" is not a perfect resistance, as defined above. Resistors are often marked with their tolerance (maximum expected variation from the marked resistance). These E numbers correspond to the formula

R = 10^(N/E), So for an 1.21 ohm E96 series resistor, N=8 and 10^(8/96)=1.21 ohm. Each multiple of 96 added to the remainder gives the next decade. So a 12.1 ohm resistor would have a N= 8+96 = 104. N can also be found by using the formula E*LOG10(R) = N.

5-band Axial Resistors

5-band identification is used for higher precision (lower tolerance) resistors (1 percent, 0.5 percent, 0.25 percent, 0.1 percent), to notate the extra digit. The first three bands represent the significant digits, the fourth is the multiplier, and the fifth is the tolerance. Five-band standard tolerance resistors are sometimes encountered, generally on older or specialized resistors. They can be identified by noting a standard tolerance color in the fourth band. The fifth band in this case is the temperature coefficient.

SMT Resistors

This image shows four surface-mount resistors (the component at the upper left is a capacitor) including two zero-ohm resistors. Zero-ohm links are often used instead of wire links, so that they can be inserted by a resistor-inserting machine.

Surface mounted resistors are printed with numerical values in a code related to that used on axial resistors. Standard-tolerance Surface Mount Technology (SMT) resistors are marked with a three-digit code, in which the first two digits are the first two significant digits of the value and the third digit is the power of ten (the number of zeroes). For example:

"334"	= 33 × 10,000 ohms = 330 kilohms
"222"	= 22 × 100 ohms = 2.2 kilohms
"473"	= 47 × 1,000 ohms = 47 kilohms
"105"	= 10 × 100,000 ohms = 1 megohm

Resistances less than 100 ohms are written: 100, 220, 470. The final zero represents ten to the power zero, which is 1. For example:

"100"	= 10 × 1 ohm = 10 ohms
"220"	= 22 × 1 ohm = 22 ohms

Sometimes these values are marked as "10" or "22" to prevent a mistake.

Resistances less than 10 ohms have 'R' to indicate the position of the decimal point (radix point). For example:

"4R7"	= 4.7 ohms
"0R22"	= 0.22 ohms
"0R01"	= 0.01 ohms

Precision resistors are marked with a four-digit code, in which the first three digits are the significant figures and the fourth is the power of ten. For example:

"1001"	= 100 × 10 ohms = 1 kilohm
"4992"	= 499 × 100 ohms = 49.9 kilohm
"1000"	= 100 × 1 ohm = 100 ohms

"000" and "0000" sometimes appear as values on surface-mount zero-ohm links, since these have (approximately) zero resistance.

Industrial Type Designation

Format: [two letters]<space>[resistance value (three digit)]<nospace>[tolerance code(numerical - one digit)]

Power Rating at 70 °C			
Type No.	**Power rating (watts)**	**MIL-R-11 Style**	**MIL-R-39008 Style**
BB	1/8	RC05	RCR05
CB	¼	RC07	RCR07
EB	½	RC20	RCR20
GB	1	RC32	RCR32
HB	2	RC42	RCR42
GM	3	-	-
HM	4	-	-

Tolerance Code		
Industrial type designation	**Tolerance**	**MIL Designation**
5	±5%	J
2	±20%	M
1	±10%	K
-	±2%	G
-	±1%	F
-	±0.5%	D
-	±0.25%	C
-	±0.1%	B

The operational temperature range distinguishes commercial grade, industrial grade and military grade components.

- Commercial grade: 0 °C to 70 °C
- Industrial grade: −40 °C to 85 °C (sometimes −25 °C to 85 °C)
- Military grade: −55 °C to 125 °C (sometimes -65 °C to 275 °C)
- Standard Grade -5°C to 60°C

Resistor Standards

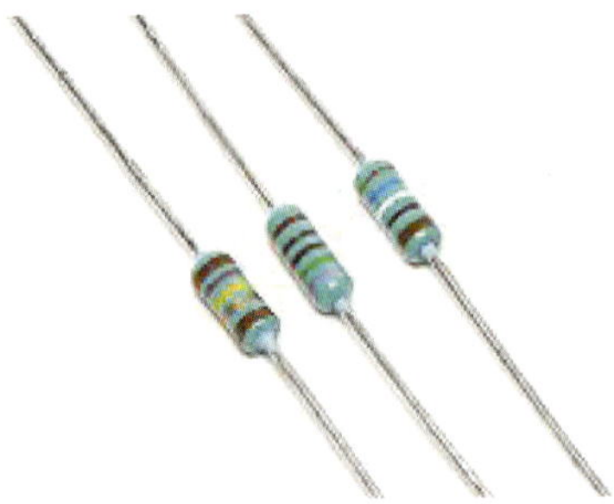

3 resistors

- MIL-R-11
- MIL-R-39008
- MIL-R-39017
- MIL-PRF-26
- MIL-PRF-39007
- MIL-PRF-55342
- MIL-PRF-914

- BS 1852
- EIA-RS-279

There are other United States military procurement MIL-R- standards.

Power Dissipation

The power dissipated by a resistor is the voltage across the resistor multiplied by the current through the resistor:

$$P = I^2 R = IV = \frac{V^2}{R}$$

All three equations are equivalent. The first is derived from Joule's law, and the other two are derived from that by Ohm's law.

The total amount of heat energy released is the integral of the power over time:

$$W = \int_{t_1}^{t_2} v(t)\, i(t)\, dt$$

If the average power dissipated exceeds the power rating of the resistor, the resistor may depart from its nominal resistance, and may be damaged by overheating. Excessive power dissipation may raise the temperature of the resistor to a point where it burns out, which could cause a fire in adjacent components and materials.

Series and Parallel Circuits

When resistors are in a parallel configuration, each one has the same potential difference (voltage) across it. To find their total equivalent resistance (R_{eq}):

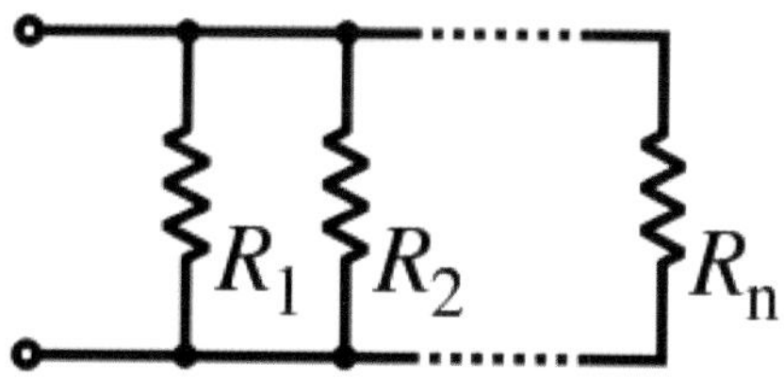

$$R_{eq} = \frac{1}{R_1} + \frac{1}{R_2} + \ldots + \frac{1}{R_n}$$

The parallel property can be represented in equations by two vertical lines "||" (as in geometry) to simplify equations. For two resistors,

$$R_{eq} = R_1 \| R_2 = \frac{R_1 R_2}{R_1 + R_2}$$

The current through resistors in series stays the same, but the voltage across each resistor can be different. The sum of the potential differences (voltage) is equal to the total voltage. To find their total resistance:

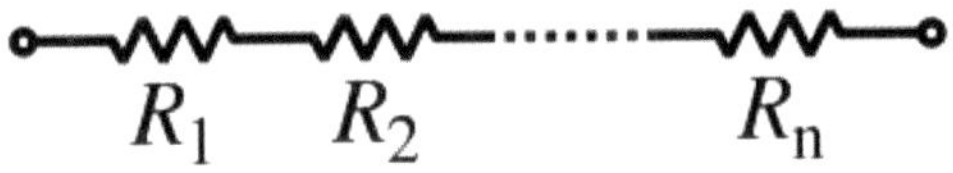

$$R_{eq} = R_1 + R_2 + \dots + R_n$$

A resistor network that is a combination of parallel and series can sometimes be broken up into smaller parts that are either one or the other. For instance,

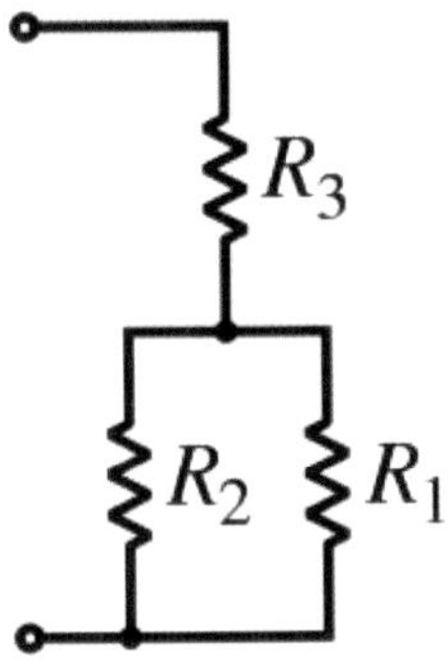

$$R_{eq} = \left(R_1 \| R_2\right) + R_3 = \frac{R_1 R_2}{R_1 + R_2} + R_3$$

However, many resistor networks cannot be split up in this way. Consider a cube, each edge of which has been replaced by a resistor. For example, determining the resistance between two opposite vertices requires matrix methods for the general case. However, if all twelve resistors are equal, the corner-to-corner resistance is $\frac{5}{6}$ of any one of them.

Technology

Carbon Composition

Carbon composition resistors consist of a solid cylindrical resistive element with embedded wire leadouts or metal end caps to which the leadout wires are attached, which is protected with paint or plastic.

The resistive element is made from a mixture of finely ground (powdered) carbon and an insulating material (usually ceramic). The mixture is held together by a resin. The resistance is determined by the ratio of the fill material (the powdered ceramic) and the carbon. Higher concentrations of carbon, a weak conductor, result in lower resistance. Carbon composition resistors were commonly used in the 1960s and earlier, but are not so popular for general use now as other types have better specifications, such as tolerance, voltage dependence, and stress (carbon composition resistors will change value when stressed with over-voltages).

Carbon Film

A spiral is used to increase the length and decrease the width of the film, which increases the resistance. Varying shapes, coupled with the resistivity of carbon, (ranging from 9 to 40 μΩm) can make for a variety of resistances.

Thick and Thin Film

Thick film resistors became popular during the 1970s, and most SMD resistors today are of this type. The principal difference between "thin film" and "thick film resistors" isn't necessarily the "thickness" of the film, but rather, how the film is applied to the cylinder (axial resistors) or the surface (SMD resistors). In thick film resistors, the "film" is applied using traditional screen-printing technology.

Thin film resistors are made by sputtering the resistive material onto the surface of the resistor. Sputtering is a method used in vacuum deposition. The thin film is then etched in a similar manner to the old (subtractive) process for making printed circuit boards: For example, the surface is coated with a photo-sensitive material, then covered by a film, irradiated with ultraviolet light, and then the exposed photo-sensitive coating, and underlying thin film, are etched away.

Thin film resistors, like their thick film counter parts are then usually trimmed to an accurate value by abrasive or laser trimming.

Because the time during which the sputtering is performed can be controlled, the thickness of the film of a thin-film resistor can be accurately controlled. The type of the material is also usually different consisting of one or more ceramic (cermet) conductors such as tantalum nitride (TaN), ruthenium dioxide (RuO_2), lead oxide (PbO), bismuth ruthenate ($Bi_2Ru_2O_7$), nickel chromium (NiCr) and bismuth iridate ($Bi_2Ir_2O_7$).

By contrast, thick film resistors, may use the same conductive ceramics, but they are mixed with sintered (powdered) glass, and some kind of liquid so that the composite can be screen-printed. This composite of glass and conductive ceramic (cermet) material is then fused (baked) in an oven at about 850 °C.

Traditionally thick film resistors had tolerances of 5 percent, but in the last few decades, standard tolerances have improved to 2 and 1 percent. But beware, temperature coefficients of thick film resistors are typically ±200 or ±250 ppm/K, depending on the resistance. Thus a 40 kelvin (70° F) temperature change can add another 1 percent variation to a 1 percent resistor.

Thin film resistors are usually specified with tolerances of 0.1, 0.2, 0.5, and 1 percent, and with temperature coefficients of 5 to 25 ppm/K. They are usually far more expensive than their thick film cousins. Note, though, that SMD thin film resistors, with 0.5 percent tolerances, and with 25 ppm/K temperature coefficients, when bought in full size reel quantities, are about twice the cost of a 1 percent, 250 ppm/K thick film resistors.

Metal Film

A common type of axial resistor today is referred to as a metal-film resistor. MELF (Metal Electrode Leadless Face) resistors often use the same technology, but are a cylindrically shaped resistor designed for surface mounting. (Note that other types of resistors, for example, carbon composition, are also available in "MELF" packages).

Metal film resistors are usually coated with nickel chromium (NiCr), but might be coated with any of the cermet materials listed above for thin film resistors. Unlike thin film resistors, the material may be applied using different techniques than sputtering (though that is one such technique). Also, unlike thin-film resistors, the resistance value is determined by cutting a helix through the coating rather than by etching. (This is similar to the way carbon resistors are made.) The result is a reasonable tolerance (0.5, 1, or 2 percent) and a temperature coefficient of (usually) 25 or 50 ppm/K.

Wirewound

Wirewound resistors are commonly made by winding a metal wire around a ceramic, plastic, or fiberglass core. The ends of the wire are soldered or welded to two caps, attached to the ends of the core. The assembly is protected with a layer of paint, molded plastic, or an enamel coating baked at high temperature. The wire leads are usually between 0.6 and 0.8 mm in diameter and tinned for ease of soldering. For higher power wirewound resistors, either a ceramic outer case or an aluminum outer case on top of an insulating layer is used. The aluminum cased types are designed to be attached to a heatsink to dissipate the heat; the rated power is dependent on being used with a suitable heatsink, for exmaple, a 50 W power rated resistor will overheat at around one fifth of the power dissipation if not used with a heatsink.

Because wirewound resistors are coils they have more inductance than other types of resistor, although this property can be minimized by winding the wire in sections with alternately reversed direction.

Foil Resistor

Foil resistors have had the best precision and stability ever since they were introduced in 1958, by Felix Zandman. One of the important parameters influencing stability is the temperature coefficient of resistance (TCR). Although the TCR of foil resistors is considered extremely low, this characteristic has been further refined over the years.

Grid Resistor

The term "grid resistor" can mean either of two things, in vacuum tube or valve electronic circuit design, a grid resistor or "grid stopper" is used to limit grid current and prevent high frequencies from entering or oscillating the circuit. Such a resistor can be composed of any one of a number of resistor technologies. One application that uses grid resistors is electric guitar amplifier circuitry.

In heavy duty, industrial, high-current applications, a grid resistor is a large convection-cooled lattice of stamped metal alloy strips connected in rows between two electrodes. Such industrial grade resistors can be as large as a refrigerator; some designs can handle over 500 amperes of current, with a range of resistances extending lower than 0.04 ohm. They are used in applications such as dynamic braking for locomotives and trams, neutral grounding for industrial AC distribution, control loads for cranes and heavy equipment, load testing of generators and harmonic filtering for electric substations.

Strain Gauges

Invented by Edward E. Simmons and Arthur C. Ruge in 1938, the strain gauge consists of a resistor that changes value with applied strain. The strain resistor may be applied singly, in pairs (half bridge), or four resistors connected in a Wheatstone bridge configuration. The strain resistor is bonded with adhesive to an object that will be subjected to the strain force. With the strain gauge and a filter, amplifier, and an analog/digital converter, it is possible to directly measure the strain on an object.

Capacitor

A capacitor is a passive electronic component that stores energy in the form of an electrostatic field. In its simplest form, a capacitor consists of two conducting plates separated by an insulating material called the dielectric. The capacitance is directly proportional to the surface areas of the plates, and is inversely proportional to the separation between the plates. Capacitance also depends on the dielectric constant of the substance separating the plates.

Capacitors can be fabricated onto integrated circuit (IC) chips. They are commonly used in conjunction with transistors in dynamic random access memory (DRAM). The capacitors help maintain the contents of memory. Because of their tiny physical size, these components have low capacitance. They must be recharged thousands of times per second or the DRAM will lose its data.

Large capacitors are used in the power supplies of electronic equipment o fall types, including computers and their peripherals. In these systems,the capacitors smooth out the rectified utility AC, providing pure, battery-like DC.

Symbols and Units

Circuit Symbols

There are two common ways to draw a capacitor in a schematic. They always have two terminals, which go on to connect to the rest of the circuit. The capacitors symbol consists of two parallel lines, which are either flat or curved; both lines should be parallel to each other, close, but not touching (this is actually representative of how the capacitor is made. Hard to describe, easier to just show.

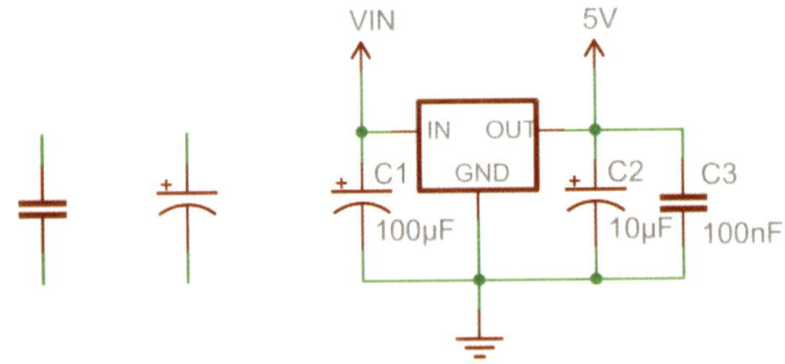

Figure: a and b are standard capacitor circuit symbols. c is an example of capacitors symbols in action in a voltage regulator circuit.

The symbol with the curved line indicates that the capacitor is polarized, meaning it's probably an electrolytic capacitor.

Each capacitor should be accompanied by a name – C1, C2, etc. – and a value. The value should indicate the capacitance of the capacitor; how many farads it has. Speaking of farads.

Capacitance Units

Not all capacitors are created equal. Each capacitor is built to have a specific amount of capacitance. The capacitance of a capacitor tells you how much charge it can store, more capacitance means more capacity to store charge. The standard unit of capacitance is called the farad, which is abbreviated *F*.

It turns out that a farad is a *lot* of capacitance, even 0.001F (1 milifarad – 1mF) is a big capacitor. Usually you'll see capacitors rated in the pico- (10^{-12}) to microfarad (10^{-6}) range.

Prefix Name	Abbreviation	Weight	Equivalent Farads
Picofarad	pF	10^{-12}	0.000000000001 F
Nanofarad	nF	10^{-9}	0.000000001 F
Microfarad	µF	10^{-6}	0.000001 F
Milifarad	mF	10^{-3}	0.001 F
Kilofarad	kF	10^{3}	1000 F

When you get into the farad to kilofarad range of capacitance, you start talking about special caps called *super* or *ultra*-capacitors.

Capacitor Theory

The stuff on this page isn't completely critical for electronics beginners to understand

and it gets a little complicated towards the end. We recommend reading the *How a Capacitor is Made* section, the others could probably be skipped if they give you a headache.

Making of a Capacitor

The schematic symbol for a capacitor actually closely resembles how it's made. A capacitor is created out of two metal plates and an insulating material called a dielectric. The metal plates are placed very close to each other, in parallel, but the dielectric sits between them to make sure they don't touch.

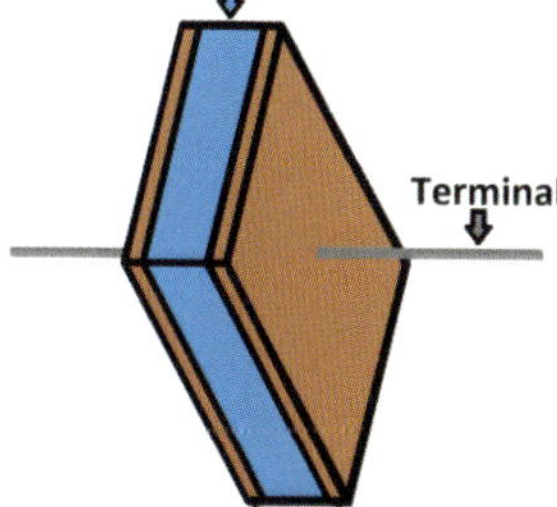

Two metal plates separated by an insulating dielectric.

The dielectric can be made out of all sorts of insulating materials: paper, glass, rubber, ceramic, plastic, or anything that will impede the flow of current.

The plates are made of a conductive material: aluminum, tantalum, silver, or other metals. They're each connected to a terminal wire, which is what eventually connects to the rest of the circuit.

The capacitance of a capacitor – how many farads it has – depends on how it's constructed. More capacitance requires a larger capacitor. Plates with more overlapping surface area provide more capacitance, while more distance between the plates means less capacitance. The material of the dielectric even has an effect on how many farads a cap has. The total capacitance of a capacitor can be calculated with the equation:

$$C = \varepsilon r \frac{A}{4\pi}$$

Where, ε_r is the dielectric's relative permittivity (a constant value determined by the dielectric material), *A* is the amount of area the plates overlap each other, and *d* is the distance between the plates.

Working of Capacitor

Electric current is the flow of electric charge, which is what electrical components harness to light up, or spin, or do whatever they do. When current flows into a capacitor, the charges get "stuck" on the plates because they can't get past the insulating dielectric. Electrons – negatively charged particles – are sucked into one of the plates, and it becomes overall negatively charged. The large mass of negative charges on one plate pushes away like charges on the other plate, making it positively charged.

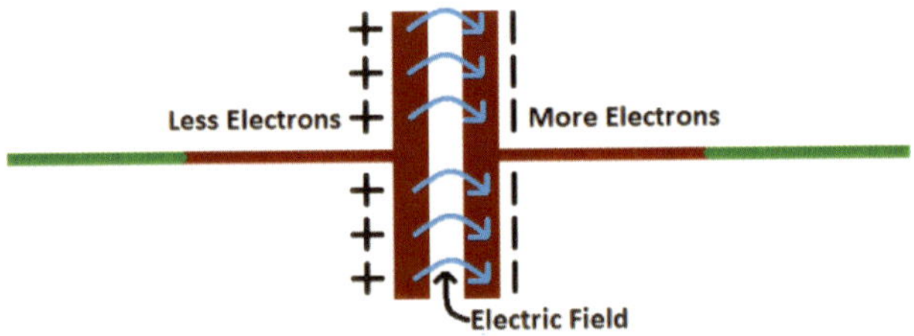

The positive and negative charges on each of these plates attract each other, because that's what opposite charges do. But, with the dielectric sitting between them, as much as they want to come together, the charges will forever be stuck on the plate (until they have somewhere else to go). The stationary charges on these plates create an electric field, which influence electric potential energy and voltage. When charges group together on a capacitor like this, the cap is storing electric energy just as a battery might store chemical energy.

Charging and Discharging

When positive and negative charges coalesce on the capacitor plates, the capacitor becomes charged. A capacitor can retain its electric field – hold its charge – because the positive and negative charges on each of the plates attract each other but never reach each other.

At some point the capacitor plates will be so full of charges that they just can't accept any more. There are enough negative charges on one plate that they can repel any others that try to join. This is where the capacitance (farads) of a capacitor comes into play, which tells you the maximum amount of charge the cap can store.

If a path in the circuit is created, which allows the charges to find another path to each other, they'll leave the capacitor, and it will discharge.

For example, in the circuit below, a battery can be used to induce an electric potential across the capacitor. This will cause equal but opposite charges to build up on each of the plates, until they're so full they repel any more current from flowing. An LED placed in series with the cap could provide a path for the current, and the energy stored in the capacitor could be used to briefly illuminate the LED.

Calculating Charge, Voltage, and Current

A capacitor's capacitance – how many farads it has – tells you how much charge it can store. How much charge a capacitor is *currently* storing depends on the potential difference (voltage) between its plates. This relationship between charge, capacitance, and voltage can be modeled with this equation:

$$Q = CV$$

Charge (Q) stored in a capacitor is the product of its capacitance (C) and the voltage (V) applied to it.

The capacitance of a capacitor should always be a constant, known value. So we can adjust voltage to increase or decrease the cap's charge. More voltage means more charge, less voltage less charge.

That equation also gives us a good way to define the value of one farad. One farad (F) is the capacity to store one unit of energy (coulombs) per every one volt.

Calculating Current

We can take the charge/voltage/capacitance equation a step further to find out how capacitance and voltage affect current, because current is the *rate* of flow of charge. The gist of a capacitor's relationship to voltage and current is this: the amount of current through a capacitor depends on both the capacitance and how quickly the voltage is rising or falling. If the voltage across a capacitor swiftly rises, a large positive current will be induced through the capacitor. A slower rise in voltage across a capacitor equates to a smaller current through it. If the voltage across a capacitor is steady and unchanging, no current will go through it.

(This is ugly, and gets into calculus. It's not all that necessary until you get into time-domain analysis, filter-design, and other gnarly stuff, so skip ahead to the next page if you're not comfortable with this equation.) The equation for calculating current through a capacitor is:

$$i = C\frac{d\mathrm{v}}{dt}$$

The *dV/dt* part of that equation is a derivative (a fancy way of saying *instantaneous rate*) of voltage over time, it's equivalent to saying "how fast is voltage going up or down at this very moment". The big takeaway from this equation is that if voltage is steady, the derivative is zero, which means current is also zero. This is why current cannot flow through a capacitor holding a steady, DC voltage.

Types of Capacitors

There are all sorts of capacitor types out there, each with certain features and drawbacks which make it better for some applications than others.

When deciding on capacitor types there are a handful of factors to consider:

- Size: Size both in terms of physical volume and capacitance. It's not uncommon for a capacitor to be the largest component in a circuit. They can also be very tiny. More capacitance typically requires a larger capacitor.

- Maximum voltage: Each capacitor is rated for a maximum voltage that can be dropped across it. Some capacitors might be rated for 1.5V, others might be rated for 100V. Exceeding the maximum voltage will usually result in destroying the capacitor.
- Leakage current: Capacitors aren't perfect. Every cap is prone to leaking some tiny amount of current through the dielectric, from one terminal to the other. This tiny current loss (usually nanoamps or less) is called leakage. Leakage causes energy stored in the capacitor to slowly, but surely drain away.
- Equivalent series resistance (ESR): The terminals of a capacitor aren't 100% conductive, they'll always have a tiny amount of resistance (usually less than 0.01Ω) to them. This resistance becomes a problem when a lot of current runs through the cap, producing heat and power loss.
- Tolerance: Capacitors also can't be made to have an exact, precise capacitance. Each cap will be rated for their nominal capacitance, but, depending on the type, the exact value might vary anywhere from ±1% to ±20% of the desired value.

Ceramic Capacitors

The most commonly used and produced capacitor out there is the ceramic capacitor. The name comes from the material from which their dielectric is made.

Ceramic capacitors are usually both physically and capacitance-wise small. It's hard to find a ceramic capacitor much larger than 10μF. A surface-mount ceramic cap is commonly found in a tiny 0402 (0.4mm x 0.2mm), 0603 (0.6mm x 0.3mm) or 0805 package. Through-hole ceramic caps usually look like small (commonly yellow or red) bulbs, with two protruding terminals.

Two caps in a through-hole, radial package; a 22pF cap on the left, and a 0.1μF on the right. In the middle, a tiny 0.1μF 0603 surface-mount cap.

Compared to the equally popular electrolytic caps, ceramics are a more near-ideal capacitor (much lower ESR and leakage currents), but their small capacitance can be limiting. They are usually the least expensive option too. These caps are well-suited for high-frequency coupling and decoupling applications.

Aluminum and Tantalum Electrolytic

Electrolytics are great because they can pack *a lot* of capacitance into a relatively small volume. If you need a capacitor in the range of 1μF-1mF, you're most likely to find it in an electrolytic form. They're especially well suited to high-voltage applications because of their relatively high maximum voltage ratings.

Aluminum electrolytic capacitors, the most popular of the electrolytic family, usually look like little tin cans, with both leads extending from the bottom.

An assortment of through-hole and surface-mount electrolytic capacitors.
Notice each has some method for marking the cathode (negative lead).

Unfortunately, electrolytic caps are usually polarized. They have a positive pin – the anode – and a negative pin called the cathode. When voltage is applied to an electrolytic cap, the anode must be at a higher voltage than the cathode. The cathode of an electrolytic capacitor is usually identified with a '-' marking, and a colored strip on the case. The leg of the anode might also be slightly longer as another indication. If voltage is applied in reverse on an electrolytic cap, they'll fail spectacularly (making a pop and bursting open), and permanently. After popping an electrolytic will behave like a short circuit.

These caps also notorious for leakage – allowing small amounts of current (on the order of nA) to run through the dielectric from one terminal to the other. This makes electrolytic caps less-than-ideal for energy storage, which is unfortunate given their high capacity and voltage rating.

Supercapacitors

If you're looking for a capacitor made to store energy, look no further than supercapacitors. These caps are uniquely designed to have *very* high capacitances, in the range of farads.

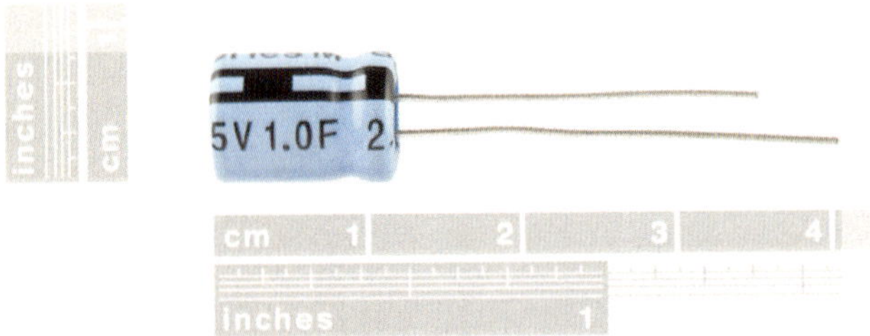

Supercapacitor: high capacitance, but only rated for 2.5V.

While they can store a huge amount of charge, supercaps can't deal with very high voltages. This 10F supercap is only rated for 2.5V max. Any more than that will destroy it. Super caps are commonly placed in series to achieve a higher voltage rating (while reducing total capacitance).

The main application for supercapacitors is in storing and releasing energy, like batteries, which are their main competition. While supercaps can't hold as much energy as an equally sized battery, they can release it much faster, and they usually have a much longer lifespan.

Others

Electrolytic and ceramic caps cover about 80% of the capacitor types out there (and supercaps only about 2%, but they're super). Another common capacitor type is the film capacitor, which features very low parasitic losses (ESR), making them great for dealing with very high currents.

There's plenty of other less common capacitors. Variable capacitors can produce a range of capacitances, which makes them a good alternative to variable resistors in tuning circuits. Twisted wires or PCBs can create capacitance (sometimes undesired) because each consists of two conductors separated by an insulator. Leyden Jars – a glass jar filled with and surrounded by conductors – are the O.G. of the capacitor family. Finally, of course, flux capacitors (a strange combination of inductor and capacitor) are critical if you ever plan on traveling back to the glory days.

Capacitors in Series/Parallel

Much like resistors, multiple capacitors can be combined in series or parallel to create a combined equivalent capacitance. Capacitors, however, add together in a way that's completely the opposite of resistors.

Capacitors in Parallel

When capacitors are placed in parallel with one another the total capacitance is simply the sum of all capacitances. This is analogous to the way resistors add when in series.

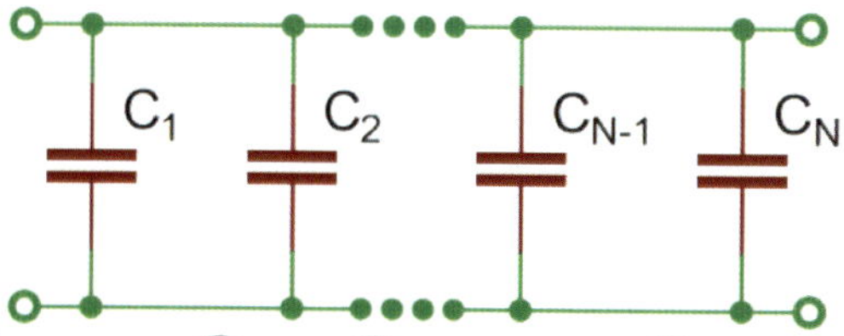

$$C_{Tot} = C_1 + C_2 + \ldots + C_{N-1} + C_N$$

So, for example, if you had three capacitors of values 10µF, 1µF, and 0.1µF in parallel, the total capacitance would be 11.1µF (10+1+0.1).

Capacitors in Series

Much like resistors are a pain to add in parallel, capacitors get funky when placed in *series*. The total capacitance of *N*-capacitors in series is the inverse of the sum of all inverse capacitances.

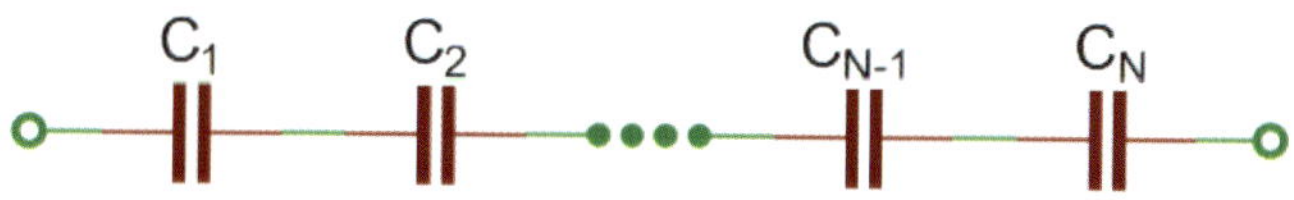

$$\frac{1}{C_{Tot}} = \frac{1}{C_1} + \frac{1}{C_2} + \ldots + \frac{1}{C_{N-1}} + \frac{1}{C_N}$$

If you only have two capacitors in series, you can use the "product-over-sum" method to calculate the total capacitance:

$$C_{Tot} = \frac{C_1 C_2}{C_1 + C_2}$$

Taking that equation even further, if you have two equal-valued capacitors in series, the total capacitance is half of their value. For example two 10F supercapacitors in series will produce a total capacitance of 5F (it'll also have the benefit of doubling the voltage rating of the total capacitor, from 2.5V to 5V).

Applications

Capacitors have various uses in electronic and electrical systems.

Energy Storage

A capacitor can store electric energy when disconnected from its charging circuit, so it can be used like a temporary battery. Capacitors are commonly used in electronic devices to maintain power supply while batteries are being changed. (This prevents loss of information in volatile memory).

Capacitor symbols		
Capacitor	Polarized capacitors	Variable capacitor
	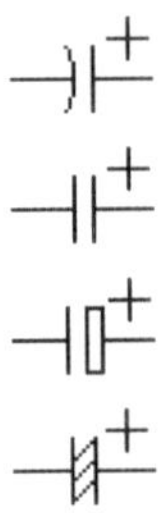	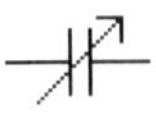

Capacitors are used in power supplies where they smooth the output of a full or half wave rectifier (device that converts AC current to DC). They can also be used in charge pump circuits as the energy storage element in the generation of higher voltages than the input voltage.

Capacitors are connected in parallel with the power circuits of most electronic devices and larger systems (such as factories) to shunt away and conceal current fluctuations from the primary power source to provide a "clean" power supply for signal or control circuits. Audio equipment, for example, uses several capacitors in this way, to shunt away power line hum before it gets into the signal circuitry. The capacitors act as a local reserve for the DC power source, and bypass AC currents from the power supply. This is used in car audio applications, when a stiffening capacitor compensates for the inductance and resistance of the leads to the lead-acid car battery.

Power Factor Correction

Capacitors are used in power factor correction to smooth out uneven current distribution. Such capacitors often come as three capacitors connected as a three phase load. Usually, the values of these capacitors are given not in farads but rather as a reactive power in volt-amperes reactive (VAr). The purpose is to counteract inductive loading from electric motors and fluorescent lighting in order to make the load appear to be mostly resistive.

Filtering

Signal Coupling

Because capacitors pass AC but block DC signals (when charged up to the applied dc voltage), they are often used to separate the AC and DC components of a signal. This method is known as *AC coupling*. (Sometimes transformers are used for the same effect.) Here, a large value of capacitance, whose value need not be accurately controlled, but whose reactance is small at the signal frequency, is employed. Capacitors for this purpose designed to be fitted through a metal panel are called feed-through capacitors, and have a slightly different schematic symbol.

Noise Filters, Motor Starters, and Snubbers

When an inductive circuit is opened, the current through the inductance collapses quickly, creating a large voltage across the open circuit of the switch or relay. If the inductance is large enough, the energy will generate a spark, causing the contact points to oxidize, deteriorate, or sometimes weld together, or destroying a solid-state switch. A snubber capacitor across the newly opened circuit creates a path for this impulse to bypass the contact points, thereby preserving their life; these were commonly found in contact breaker ignition systems, for instance. Similarly, in smaller scale circuits, the spark may not be enough to damage the switch but will still radiate undesirable radio frequency interference (RFI), which a filter capacitor absorbs. Snubber capacitors are usually employed with a low-value resistor in series, to dissipate energy and minimize RFI. Such resistor-capacitor combinations are available in a single package.

In an inverse fashion, to initiate current quickly through an inductive circuit requires a greater voltage than required to maintain it; in uses such as large motors, this can cause undesirable startup characteristics, and a motor starting capacitor is used to increase the coil current to help start the motor.

Capacitors are also used in parallel to interrupt units of a high-voltage circuit breaker in order to equally distribute the voltage between these units. In this case they are called grading capacitors.

In schematic diagrams, a capacitor used primarily for DC charge storage is often drawn vertically in circuit diagrams with the lower, more negative, plate drawn as an arc. The straight plate indicates the positive terminal of the device, if it is polarized.

Signal Processing

The energy stored in a capacitor can be used to represent information, either in binary form, as in DRAMs, or in analogue form, as in analog sampled filters and CCDs. Capacitors can be used in analog circuits as components of integrators or more complex filters and in negative feedback loop stabilization. Signal processing circuits also use capacitors to integrate a current signal.

Tuned Circuits

Capacitors and inductors are applied together in tuned circuits to select information in particular frequency bands. For example, radio receivers rely on variable capacitors to tune the station frequency. Speakers use passive analog crossovers, and analog equalizers use capacitors to select different audio bands.

In a tuned circuit such as a radio receiver, the frequency selected is a function of the inductance (L) and the capacitance (C) in series, and is given by:

$$f = \frac{1}{2\pi\sqrt{LC}}$$

This is the frequency at which resonance occurs in an LC circuit.

Other Applications

Sensing

Most capacitors are designed to maintain a fixed physical structure. However, various things can change the structure of the capacitor—the resulting change in capacitance can be used to sense those things.

Changing the dielectric: The effects of varying the physical and/or electrical characteristics of the dielectric can also be of use. Capacitors with an exposed and porous dielectric can be used to measure humidity in air.

Changing the distance between the plates: Capacitors are used to accurately measure the fuel level in airplanes. Capacitors with a flexible plate can be used to measure strain or pressure. Capacitors are used as the sensor in condenser microphones, where one plate is moved by air pressure, relative to the fixed position of the other plate. Some accelerometers use MEMS capacitors etched on a chip to measure the magnitude and direction of the acceleration vector. They are used to detect changes in acceleration, e.g., as tilt sensors or to detect free fall, as sensors triggering airbag deployment, and in many other applications. They are also used in fingerprint sensors.

Pulsed Power and Weapons

Groups of large, specially constructed, low-inductance high-voltage capacitors *(capacitor banks)* are used to supply huge pulses of current for many pulsed power applications. These include electromagnetic forming, Marx generator, pulsed lasers (especially TEA lasers), pulse forming networks, radar, fusion research, and particle accelerators.

Large capacitor banks are used as energy sources for the exploding-bridgewire detonators or slapper detonators in nuclear weapons and other specialty weapons. Experimental work is under way using banks of capacitors as power sources for electromagnetic armour and electromagnetic railguns or coilguns.

Hazards and Safety

Capacitors may retain a charge long after power is removed from a circuit; this charge can cause shocks (sometimes fatal) or damage to connected equipment. For example, even a seemingly innocuous device such as a disposable camera flash unit powered by a

1.5 volt AA battery contains a capacitor which may be charged to over 300 volts. This is easily capable of delivering an extremely painful, and possibly lethal shock.

Care must be taken to ensure that any large or high-voltage capacitor is properly discharged before servicing the containing equipment. For safety purposes, all large capacitors should be discharged before handling. For board-level capacitors, this is done by placing a bleeder resistor across the terminals, whose resistance is large enough that the leakage current will not affect the circuit, but small enough to discharge the capacitor shortly after power is removed. High-voltage capacitors should be stored with the terminals shorted, since temporarily discharged capacitors can develop potentially dangerous voltages when the terminals are left open-circuited.

Large oil-filled old capacitors must be disposed of properly as some contain polychlorinated biphenyls (PCBs). It is known that waste PCBs can leak into groundwater under landfills. If consumed by drinking contaminated water, PCBs are carcinogenic, even in very tiny amounts. If the capacitor is physically large it is more likely to be dangerous and may require precautions in addition to those described above. New electrical components are no longer produced with PCBs. ("PCB" in electronics usually means printed circuit board, but the above usage is an exception.) Capacitors containing PCB were labelled as containing "Askarel" and several other trade names.

High-voltage

Above and beyond usual hazards associated with working with high voltage, high energy circuits, there are a number of dangers that are specific to high voltage capacitors. High voltage capacitors may catastrophically fail when subjected to voltages or currents beyond their rating, or as they reach their normal end of life. Dielectric or metal interconnection failures may create arcing within oil-filled units that vaporizes dielectric fluid, resulting in case bulging, rupture, or even an explosion that disperses flammable oil, starts fires, and damages nearby equipment. Rigid cased cylindrical glass or plastic cases are more prone to explosive rupture than rectangular cases due to an inability to easily expand under pressure. Capacitors used in RF or sustained high current applications can overheat, especially in the center of the capacitor rolls. The trapped heat may cause rapid interior heating and destruction, even though the outer case remains relatively cool. Capacitors used within high energy capacitor banks can violently explode when a fault in one capacitor causes sudden dumping of energy stored in the rest of the bank into the failing unit. And, high voltage vacuum capacitors can generate soft X-rays even during normal operation. Proper containment, fusing, and preventative maintenance can help to minimize these hazards.

High voltage capacitors can benefit from a pre-charge to limit in-rush currents at power-up of HVDC circuits. This will extend the life of the component and may mitigate high voltage hazards.

Inductor

An inductor is a passive electrical component that can store energy in a magnetic field created by passing an electric current through it. A simple inductor is a coil of wire. When an electric current is passed through the coil, a magnetic field is formed around it. This magnetic field causes the inductor to resist changes in the amount of current passing through it.

An inductor's ability to store magnetic energy is measured by its inductance, in units of henries. The inductance of a coil is directly proportional to the number of turns in the coil. Inductance also varies with the coil's radius and the material (or "core") around which the coil is wound.

Inductors are widely used in analog circuits and signal processing. Large inductors, in combination with capacitors, are useful as chokes in power supplies, to remove fluctuations from direct current output. Small inductor/capacitor combinations are useful in making tuned circuits for radio reception and broadcasting. In addition, inductors are employed in transformers for power grids, and as energy storage devices in some switched-mode power supplies.

Overview

When an electric current first begins to pass through an inductor (coil of wire), the inductor resists the flow of current, as a magnetic field is building up around the inductor. After this field has been built up, the inductor allows current to flow through it normally. When the electricity supply is turned off, the magnetic field around the coil supports the flow of current for a brief interval, before the field disappears.

Inductance *(L)* (measured in henries) is an effect resulting from the magnetic field that forms around a current-carrying conductor that tends to resist changes in the current. Electric current through the conductor creates a magnetic flux proportional to the current. A change in this current creates a change in magnetic flux that, in turn, by Faraday's law generates an electromotive force (EMF) that acts to oppose this change in current. Inductance is a measure of the amount of EMF generated for a unit change in current. For example, an inductor with an inductance of 1 henry produces an EMF of 1 volt when the current through the inductor changes at the rate of 1 ampere per second.

The number of loops, the size of each loop, and the material it is wrapped around all affect the inductance. For example, the magnetic flux linking these turns can be increased by coiling the conductor around a material with a high permeability such as iron. This can increase the inductance by 2000 times, although less so at high frequencies.

An "ideal inductor" has inductance but no resistance or capacitance, and it does not dissipate energy. A real inductor has not only inductance but also some resistance (due to resistivity of the wire) and some capacitance. At some frequency, usually much higher than the working frequency, a real inductor behaves as a resonant circuit (due to its self capacitance). In addition to dissipating energy in the resistance of the wire, magnetic core inductors may dissipate energy in the core due to hysteresis, and at high currents may show other departures from ideal behavior due to nonlinearity.

Hydraulic Analogy

The behavior of an inductor can be described by using a hydraulic analogy. An inductor can be modeled by the flywheel effect of a heavy turbine rotated by the flow. When water (current) first starts to flow, the stationary turbine will cause an obstruction in the flow and high pressure (voltage) opposing the flow until it gets turning. Once it is turning, if there is a sudden interruption of water flow, the turbine will continue to turn by inertia, generating a high pressure to keep the flow moving. (Magnetic interactions in transformers are not modeled hydraulically).

Applications

Inductors are used extensively in analog circuits and signal processing. Along with capacitors and other components, inductors are used to form tuned circuits that can emphasize or filter out specific signal frequencies. This can range from the use of large inductors as chokes in power supplies, which in conjunction with filter capacitors remove residual hum or other fluctuations from the direct current output, to such small inductances as generated by a ferrite bead or torus around a cable to prevent radio frequency interference from being transmitted down the wire. Smaller inductor/capacitor combinations provide tuned circuits used in radio reception and broadcasting, for instance.

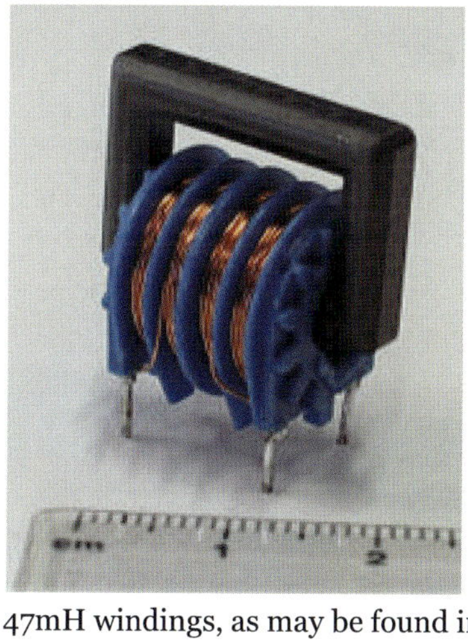

A choke with two 47mH windings, as may be found in a power supply.

Two (or more) inductors that have coupled magnetic flux form a transformer, which is a fundamental component of every electric utility power grid. The efficiency of a transformer may decrease as the frequency increases due to eddy currents in the core material and skin effect on the windings. Size of the core can be decreased at higher frequencies and, for this reason, aircraft use 400 hertz alternating current rather than the usual 50 or 60 hertz, greatly saving on weight through the use of smaller transformers.

An inductor is used as the energy storage device in some switched-mode power supplies. The inductor is energized for a specific fraction of the regulator's switching frequency, and de-energized for the remainder of the cycle. This energy transfer ratio determines the input-voltage to output-voltage ratio. This X_L is used in complement with an active semiconductor device to maintain very accurate voltage control.

Inductors are also employed in electrical transmission systems, where they are used to depress voltages from lightning strikes and to limit switching currents and fault current. In this field, they are more commonly referred to as reactors.

As inductors tend to be larger and heavier than other components, their use has been reduced in modern equipment; solid-state switching power supplies eliminate large transformers, for instance, and circuits are designed to use only small inductors, if any; larger values are simulated by use of gyrator circuits.

Inductor Construction

Inductors. (Major scale shown in centimeters.)

An inductor is usually constructed as a coil of conducting material, typically copper wire, wrapped around a core either of air or of ferromagnetic material. Core materials with a higher permeability than air increase the magnetic field and confine it closely to the inductor, thereby increasing the inductance. Low frequency inductors are constructed like transformers, with cores of electrical steel laminated to prevent eddy currents. "Soft" ferrites are widely used for cores above audio frequencies, since they don't cause the large energy losses at high frequencies that ordinary iron alloys do. This is because of their narrow hysteresis curves, and their high resistivity prevents eddy currents. Inductors come in many shapes. Most are constructed as enamel coated wire wrapped around a ferrite bobbin with wire exposed on the outside, while some enclose the wire completely in ferrite and are called "shielded." Some inductors have an adjustable core, which enables changing of the inductance. Inductors used to block very high frequencies are sometimes made by stringing a ferrite cylinder or bead on a wire.

Small inductors can be etched directly onto a printed circuit board by laying out the trace in a spiral pattern. Some such planar inductors use a planar core.

Small value inductors can also be built on integrated circuits using the same processes that are used to make transistors. Aluminum interconnect is typically used, laid out in a spiral coil pattern. However, the small dimensions limit the inductance, and it is far more common to use a circuit called a "gyrator" which uses a capacitor and active components to behave similarly to an inductor.

Calculations for Electric Circuits

An inductor opposes changes in current. An ideal inductor would offer no resistance to a constant direct current; however, only superconducting inductors have truly zero electrical resistance.

In general, the relationship between the time-varying voltage $v(t)$ across an inductor with inductance L and the time-varying current $i'(t)$ passing through it is described by the differential equation:

$$v(t) = L\frac{di(t)}{dt}$$

When, there is a sinusoidal alternating current (AC) through an inductor, a sinusoidal voltage is induced. The amplitude of the voltage is proportional to the product of the amplitude (I_p) of the current and the frequency (f) of the current.

$$i(t) = I_p \sin(2\pi ft)$$

$$\frac{di(t)}{dt} = 2\pi f I_p \cos(2\pi ft)$$

$$v(t) = 2\pi f L I_p \cos(2\pi ft)$$

In this situation, the phase of the current lags that of the voltage by 90 degrees. #

If an inductor is connected to a DC current source, with value I via a resistance, R, and then the current source short circuited, the differential relationship above shows that the current through the inductor will discharge with an exponential decay:

$$i(t) = I(e^{\frac{-tR}{L}})$$

Laplace Circuit Analysis (S-domain)

When using the Laplace transform in circuit analysis, the transfer impedance of an ideal inductor with no initial current is represented in the s domain by:

$$Z(s) = Ls$$

where,

L is the inductance, and

s is the complex frequency

If the inductor does have initial current, it can be represented by:

- adding a voltage source in series with the inductor, having the value:

$$LI_0$$

(Note that the source should have a polarity that opposes the initial current)

- or by adding a current source in parallel with the inductor, having the value:

$$\frac{I_0}{s}$$

where,

L is the inductance, and

I_0 *is the initial current in the inductor.*

Inductor Networks

Inductors in a parallel configuration each have the same potential difference (voltage). To find their total equivalent inductance (L_{eq}):

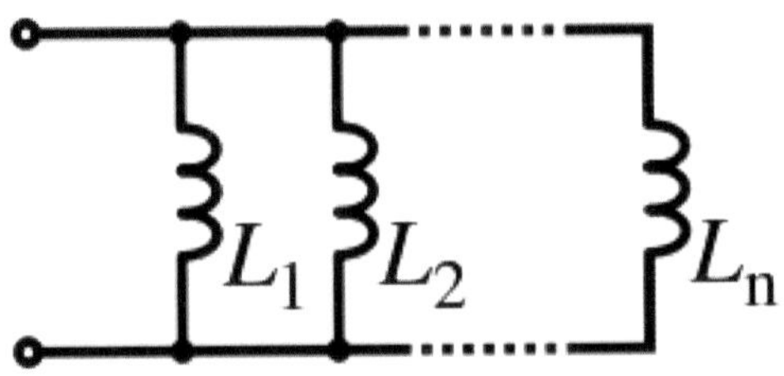

$$\frac{1}{L_{eq}} = \frac{1}{L_1} + \frac{1}{L_2} + \cdots + \frac{1}{L_n}$$

The current through inductors in series stays the same, but the voltage across each inductor can be different. The sum of the potential differences (voltage) is equal to the total voltage. To find their total inductance:

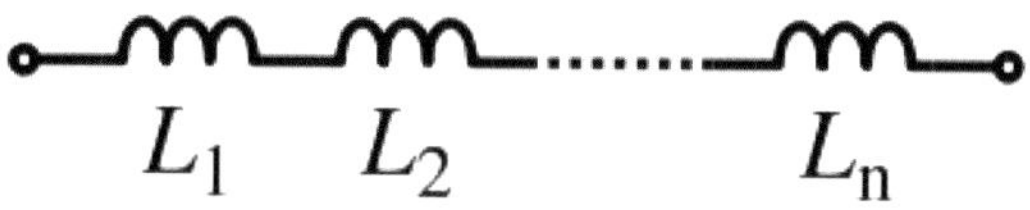

$$L_{eq} = L_1 + L_2 + \cdots + L_n$$

These simple relationships hold true only when there is no mutual coupling of magnetic fields between individual inductors.

Stored Energy

The energy (measured in joules, in SI) stored by an inductor is equal to the amount of work required to establish the current through the inductor, and therefore the magnetic field. This is given by:

$$E_{stored} = \frac{1}{2} L I^2$$

where, L is inductance and I is the current through the inductor (****).

Q factor

An ideal inductor will be lossless, irrespective of the amount of current through the winding. However, typically inductors have winding resistance from the metal wire forming the coils. Since the winding resistance appears as a resistance in series with the inductor, it is often called the *series resistance*. The inductor's series resistance converts electrical current through the coils into heat, thus causing a loss of inductive quality. The quality factor (or *Q*) of an inductor is the ratio of its inductive reactance to its resistance at a given frequency, and is a measure of its efficiency. The higher the Q factor of the inductor, the closer it approaches the behavior of an ideal, loss less, inductor.

The Q factor of an inductor can be found through the following formula, where R is its internal electrical resistance and ωL is capacitive or inductive reactance at resonance:

$$Q = \frac{\omega L}{R}$$

By using a ferromagnetic core, the inductance is greatly increased for the same amount of copper, multiplying up the Q. Cores however also introduce losses that increase with frequency. A grade of core material is chosen for best results for the frequency band. At VHF or higher frequencies an air core is likely to be used.

Inductors wound around a ferromagnetic core may saturate at high currents, causing a dramatic decrease in inductance (and Q). This phenomenon can be avoided by using a (physically larger) air core inductor. A well designed air core inductor may have a Q of several hundred.

An almost ideal inductor (Q approaching infinity) can be created by immersing a coil made from a superconducting alloy in liquid helium or liquid nitrogen. This supercools the wire, causing its winding resistance to disappear. Because a superconducting inductor is virtually lossless, it can store a large amount of electrical energy within the surrounding magnetic field.

Inductance Formulae

The table below lists some common formulae for calculating the theoretical inductance of several inductor constructions.

Construction	Formula	Dimensions
Cylindrical coil	$L = \frac{\mu_0 K N^2 A}{l}$	• L = inductance in henries (H) • μ_o = permeability of free space = 4π × 10^{-7} H/m • K = Nagaoka coefficient • N = number of turns • A = area of cross-section of the coil in square metres (m^2) • l = length of coil in metres (m)
Straight wire conductor	$L = l\left(\ln\frac{4l}{d} - 1\right)\cdot 200 \times 10^{-9}$	• L = inductance (H) • l = length of conductor (m) • d = diameter of conductor (m)
	$L = 5.08 \cdot l\left(\ln\frac{4l}{d} - 1\right)$	• L = inductance (nH) • l = length of conductor (in) • d = diameter of conductor (in)
Short air-core cylindrical coil	$L = \frac{r^2 N^2}{9r + 10l}$	• L = inductance (μH) • r = outer radius of coil (in) • l = length of coil (in) • N = number of turns
Multilayer air-core coil	$L = \frac{0.8 r^2 N^2}{6r + 9l + 10d}$	• L = inductance (μH) • r = mean radius of coil (in) • l = physical length of coil winding (in) • N = number of turns • d = depth of coil (outer radius minus inner radius) (in)

Flat spiral air-core coil	$$L=\frac{r^2N^2}{(2r+2.8d)\times 10^5}$$	• L = inductance (H) • r = mean radius of coil (m) • N = number of turns • d = depth of coil (outer radius minus inner radius) (m)
	$$L=\frac{r^2N^2}{8r+11d}$$	• L = inductance (μH) • r = mean radius of coil (in) • N = number of turns • d = depth of coil (outer radius minus inner radius) (in)
Toroidal core (circular cross-section)	$$L=\mu_0\mu_r\frac{N^2r^2}{D}$$	• L = inductance (H) • μ_o = permeability of free space = $4\pi \times 10^{-7}$ H/m • μ_r = relative permeability of core material • N = number of turns • r = radius of coil winding (m) • D = overall diameter of toroid (m)

References

- Electronic-components: electronicscoach.com, Retrieved 16 March 2018
- Transistors-basics-types-baising-modes: elprocus.com, Retrieved 26 June 2018
- What-is-a-battery, tutorials: learn.sparkfun.com, Retrieved 11 July 2018
- What-is-a-resistor: resistorguide.com, Retrieved 11 April 2018
- Capacitor-capacitance: techtarget.com, Retrieved 21 May 2018

DC and AC Circuits

An electronic circuit is a system of electronic components connected by electrical wires for the flow of electricity. The primary electronic circuits are AC and DC circuits. All the diverse aspects of AC and DC circuits, such as resistance, impedance, phasors, transfer functions, etc. have been extensively detailed in this chapter.

DC Circuit

The closed path in which the direct current flows is called the DC circuit. The current flows in only one direction and it is mostly used in low voltage applications. The resistor is the main component of the DC circuit.

A simple DC circuit is shown in the figure below which contains a DC source (battery), a load lamp, a switch, connecting leads, and measuring instruments like ammeter and voltmeter. The load resistor is connected in series, parallel or series-parallel combination as per requirement.

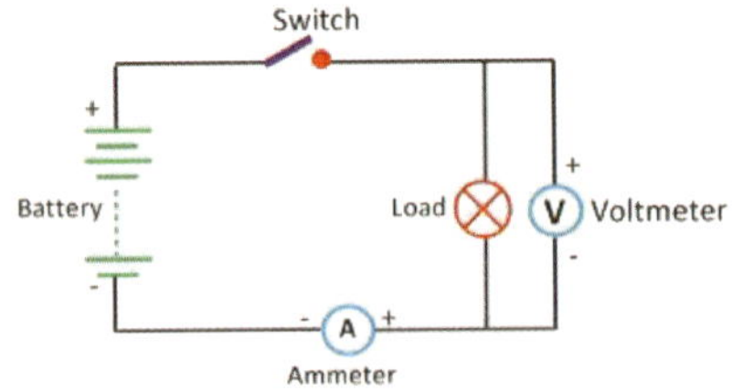

DC Circuit

Types of DC Circuit

The DC electric circuit is mainly classified into three groups. They are the series DC circuit, parallel DC circuit, and series and parallel DC circuit.

DC Series Circuit

The circuit in which have DC series source, and the number of resistors are connected end to end so that same current flow through them is called a DC series circuit. The figure below shows the simple series circuit. In the series circuit the resistor R_1, R_2, and R_3 are connected in series across a supply voltage of V volts. The same current I is flowing through all the three resistors.

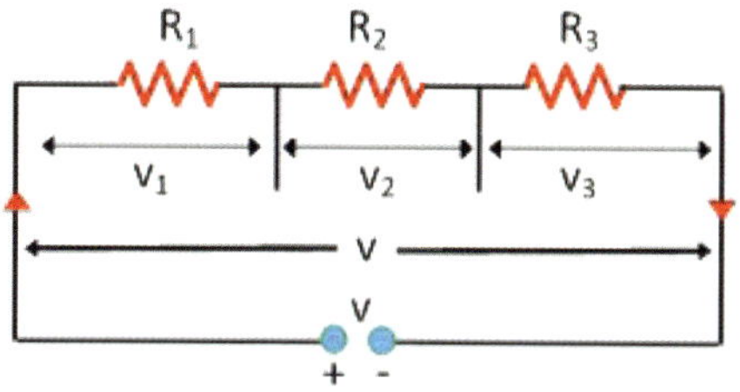

DC Series Circuit

If V_1, V_2, and V_3 are the voltage drop across the three resistor R_1, R_2, and R_3 respectively, then:

$$V = V_1 + V_2 + V_3$$
$$V = IR_1 + IR_2 + IR_3$$

Let R be the total resistance of the circuit then:

$$IR = IR_1 + IR_2 + IR_3$$
$$R = R_1 + R_2 + R_3$$

Total resistance = Sum of the individual resistance.

In such type of circuit all the lamps are controlled by the single switch and they cannot be controlled individually. The most common application of this circuit is for decoration purpose where a number of low voltage lamps are connected in series.

DC Parallel Circuit

The circuit which have DC source and one end of all the resistors is joined to a common point and other end are also joined to another common point so that current flows through them is called a DC parallel circuit.

The figure shows a simple parallel circuit. In this circuit the three resistor R_1, R_2, and R_3 are connected in parallel across a supply voltage of V volts. The current flowing through them is I_1, I_2 and I_3 respectively.

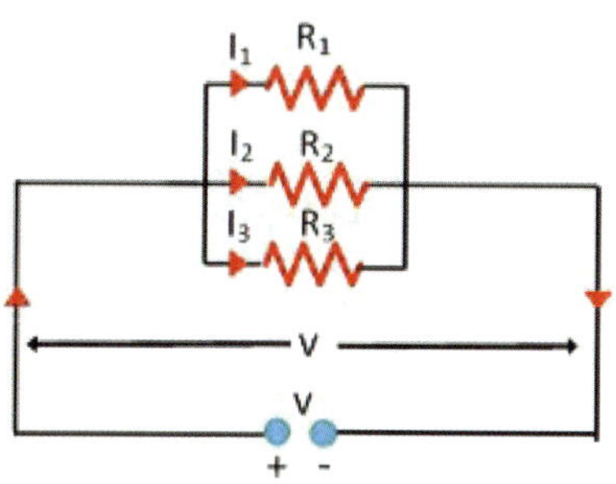

DC Parallel Circuit

The total current drawn by the circuit:

$$I = I_1 + I_2 + I_3$$

$$I = \frac{V}{R_1} + \frac{V}{R_2} + \frac{V}{R_3}$$

Let R be the total or effective resistance of the circuit, then:

$$\frac{V}{R} = \frac{V}{R_1} + \frac{V}{R_2} + \frac{V}{R_3}$$

$$\frac{1}{R} = \frac{1}{R_1} + \frac{1}{R_2} + \frac{1}{R_3}$$

Reciprocal of total resistance = sum of reciprocal of the individual resistance.

All the resistance is operated to the same voltage, therefore all of them are connected in parallel. Each of them can be controlled individually with the help of a separate switch.

DC Series-Parallel Circuit

The circuit in which series and parallel circuit are connected in series is called a series parallel circuit.The figure below show the series-parallel circuit. In this circuit, two resistor R_1 and R_2 are connected in parallel with each other across terminal AB. The other three resistors R_3, R_4 and R_6 are connected in parallel with each other across terminal BC.

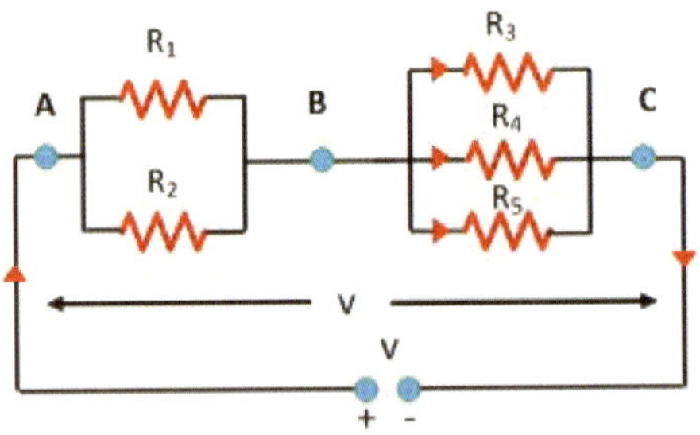

DC Series Parallel Circuit

The two groups of resistor R_{AB} and R_{BC} are connected in series with each other across the supply voltage of V volts. The total or effective resistance of the whole circuit can be determined as given below:

$$\frac{1}{R_{AB}} = \frac{1}{R_1} + \frac{1}{R_2} = \frac{R_1 + R_2}{R_1 R_2}$$

$$R_{AB} = \frac{R_1 R_2}{R_1 + R_2}$$

Similarly,

$$\frac{1}{R_{BC}} = \frac{1}{R_3} + \frac{1}{R_4} + \frac{1}{R_5} = \frac{R_3 R_4 + R_4 R_5 + R_5 R_3}{R_3 R_4 R_5}$$

$$R_{BC} = \frac{R_3 R_4 R_5}{R_3 R_4 + R_4 R_5 + R_5 R_3}$$

Total or effective resistance of the ciruit:

$$R = R_{AB} + R_{BC}$$

An Example of Parallel DC Circuit

Suppose three resistors R_1, R_2 and R_3 are connected in parallel across a voltage source of V (volt) as shown in the figure. Let I (Ampere) be the total circuit current which is divided into current I_1, I_2 and I_3 flowing through R_1, R_2 and R_3 respectively. Now according to Ohm's law: Voltage drop across resistor R_1, $V = I_1.R_1$ Voltage drop across resistor R_2, $V = I_2.R_2$ Voltage drop across resistor R_3, $V = I_3.R_3$ Voltage drop across whole parallel DC circuit, V = Voltage drop across resistor R_1 = voltage drop across resistor R_2 = voltage drop across resistor $R_3 \Rightarrow V = I_1.R_1 = I_2.R_2 = I_3.R_3$

$$Therefore,\ I_1 = \frac{V}{R_1},\ I_2 = \frac{V}{R_2},\ and\ I_3 = \frac{V}{R_3}$$

$$I = I_1 + I_2 + I_3 \ and\ as\ per\ Ohm's\ law,\ I = \frac{V}{R}\ hence,$$

$$Therefore,\ I = \frac{V}{R} = I_1 + I_2 + I_3 = \frac{V}{R_1} = \frac{V}{R_2} = \frac{V}{R_3} = V\left[\frac{1}{R_1} + \frac{1}{R_2} + \frac{1}{R_3}\right]$$

$$Therefore\ from\ above\ expression\ we\ get,\ \frac{1}{R} = \frac{1}{R_1} + \frac{1}{R_2} + \frac{1}{R_3}$$

Thus when a number of resistors are connected in parallel, the reciprocal of the equivalent resistance is given by the arithmetic sum of the reciprocals of their individual resistances. From the above discussion of parallel DC circuit, we can come to the following conclusion:

- Voltage drops are the same across all the components connected in parallel.
- Current through individual components connected in parallel, is inversely proportional to their resistances.
- Total circuit current is the arithmetic sum of the currents passing through individual components connected in parallel.

- The reciprocal of equivalent resistance is equal to the sum of the reciprocals of the resistances of individual components connected in parallel.

DC Circuit Parameters

Electric Voltage

The potential difference between two points or voltage in an electric circuit is the amount of energy required to move a unit charge between two points. It is measured in Volts and indicated with a letter V as shown in below figure. This voltage can be either positive or negative and expressed mostly with prefixes like KV, mV, uV, etc. to denote sub-multiples of the voltage. Batteries and generators are the most commonly used DC voltage sources which can produce the DC voltage from 1V to 24V DC for functioning of general electronic circuits.

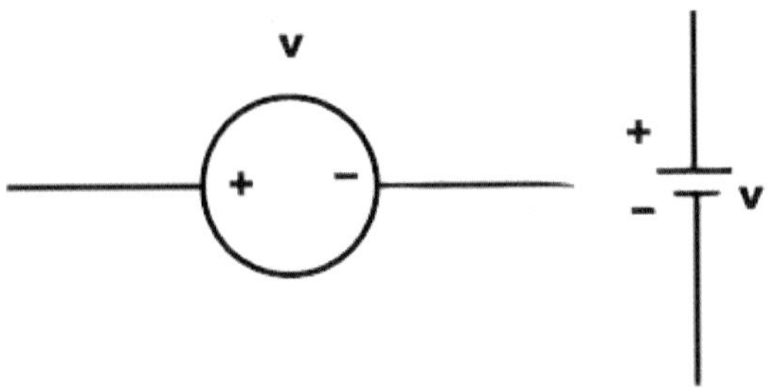

Electric Current

It is the flow of electrons or electric charge. It is measured in Amperes or simply Amps, and denoted by the letter 'I' or lower case i. This electric current can be direct or alternating. The Direct Current (DC) flows in a unidirectional way and generally it is produced by batteries, solar cells, thermocouples, etc. In case of AC, electric charge movement periodically changes as we can observe in case of sine wave.

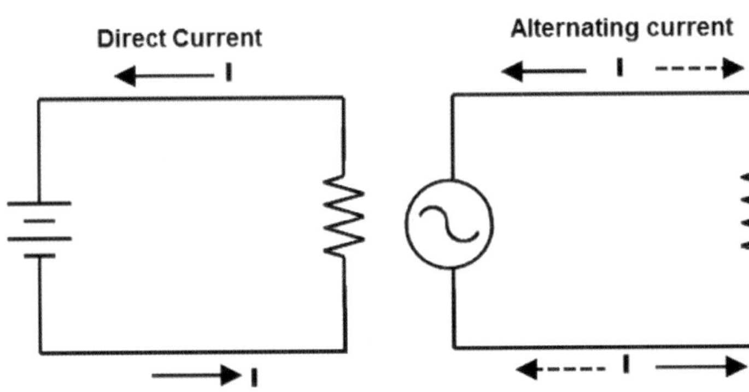

Generally in circuits the direction of current flow is indicated with a letter I or lower case I with an arrow associated with it. But this direction actually indicates the conventional current flow rather than actual electron current flow.

Difference between Conventional and Electron Current Flow

Electrons flow from negative terminal to positive terminal is referred as electron current flow, whereas from positive terminal to the negative terminal is referred as conventional current flow as shown in figure.

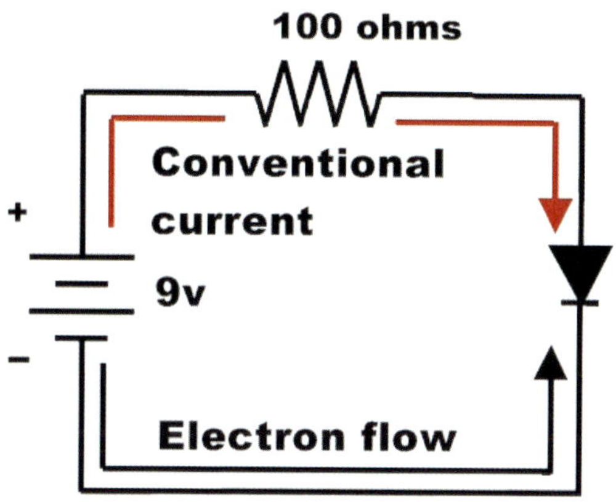

The electrons have always been repelled by the negative charge where the terminal is connected to the negative terminal of the battery and are attracted at positive terminal due to the positive charge. Hence the electrons flow from negative terminal to positive terminal is referred as electron current flow. But conventional method of assuming current flow is from positive to negative so this is referred as conventional current flow. Conventional current is indicated on many circuit diagrams and actual electron flow current is indicated in the case of describing the individual current flow.

The conventional current flow is due to the positive charge carriers. The conventional current is measured in the opposite direction of actual electron current flow, which is due to the negative charge carriers (Electrons) therefore, conventional current is always positive. It is also measured in Amps.

The difference of conventional and actual electron flow does not effect on any computational results and real time behavior. Most of the analyzing concepts of DC circuit results are independent of the direction of current flow. However, the conventional current is the standard and mostly follows.

Resistance

The resistance of a conducting material opposes the flow of electrons. It is measured in ohms and denoted by the Greek symbol Ω. Depends on the resistor value in a circuit voltage applied to the circuit is decided. Thus, resistance can be defined as the voltage required for a circuit for making 1 ampere current flow. This also referred as Ohm's law and written as $R = V/I$. That means if a circuit requires 200V to produce 2A current then the resistance should be 100 ohms. The resistance value is always positive. Resistors can be fixed or variable resistors as shown in figure:

Electric Power (P) and Energy

The power is termed as the work done in a given amount of time. In electrical circuits, power is exactly equal to the product of voltage and current. Since the voltage is the work per unit charge and current is the rate at which electrons move in a circuit. The Power is measured in watts (W) and its formula is:

$$P = I \times V$$

According to Ohms law,

$$R = V/I$$

$$V = IR$$

Substituting in the above equation,

$$P = (IR)\ R$$

$$P = I^2R$$

Or also, by substituting $I = V/R$, we can get

$$P = V \times (V/R)$$

$$P = V^2/R$$

These three possible formulas are used for finding the power associated with a circuit.

Electrical Energy

The rate at which electrical power consumed is generally referred as electrical energy. The energy is measured in watt-seconds as the power measured in watts and time in seconds. Often it is measured in kilowatt-hours as we can observe in our home electricity meter.

Electrical energy = power × time

DC Circuit Analysis

A network of passive elements and sources is a circuit.

Analysis: To determine currents or voltages in various elements (effects) due to various sources (cause).

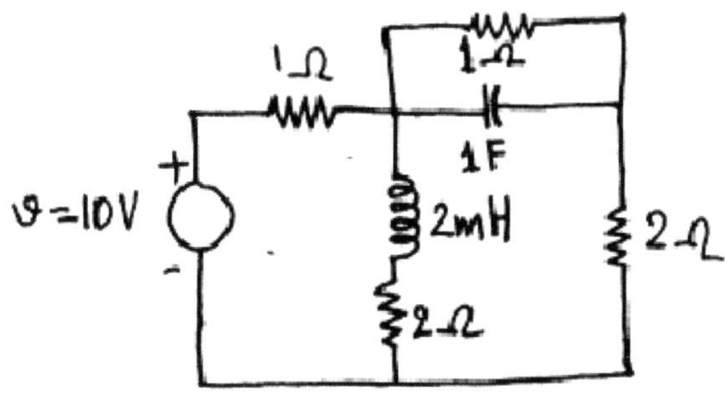

In circuit $v = 10V$ all the time.

Expected that all current (voltage) in (across) the elements is constant.

Inductor

$$v = L\frac{di}{dt}$$

$$\frac{di}{dt} = 0 \text{ implies } i - \text{constant.}$$

$$\text{Hence } v = 0$$

Inductors act as a short cicuit for DC inputs. This would not be the case if I put a switch across a source.

Capacitor

$$i = C\frac{dv}{dt}$$

$$\text{as } \frac{dv}{dt} = 0 \text{ (expected), } i = 0.$$

Thus capacitor acts as open circuit for DC analysis.

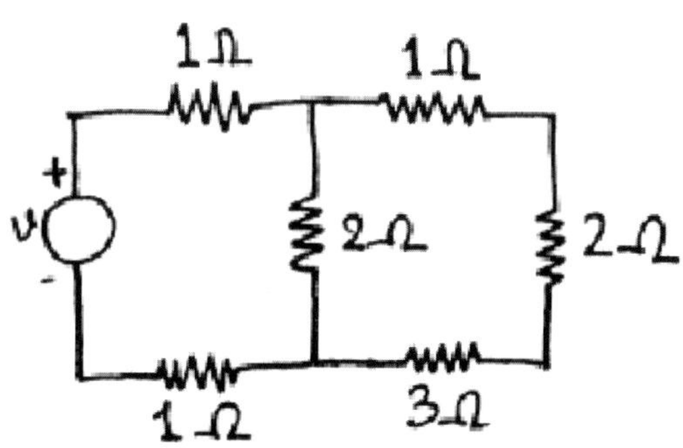

The resultant circuit will be as shown in figure.

Analysis: To find currents in all branches, voltage across all branches.

We can use Kirchoff's law (voltage and current). For as many independent equations as number of unknown variables.

Solve the simultaneous equations, and get the result.

V_{ab} = Voltage drop from 'a' to 'b'. Therefore, $V_{ab} = -V_{ba}$

I_{ab} = Current in branch ab in the direction from 'a' to 'b'. Here $I_{ab} = -I_{ba}$.

Hence:

$$V_{fa} = -v$$
$$V_{ab} = I_{ab} \times 1 = I_{ab}$$
$$V_{be} = I_{be} 2$$
$$V_{ef} = I_{ef} 1$$
$$V_{bc} = I_{bc} 1$$
$$V_{cd} = I_{de} 2$$
$$V_{de} = I_{de} 3$$
$$V_{ab} + V_{be} + V_{ef} = v$$
$$I_{fa} = I_{ab}$$
$$I_{ef} = I_{fa}$$
$$I_{bc} = I_{cd}$$
$$I_{cd} = I_{de}$$
$$V_{ab} + V_{bc} + V_{cd} + V_{de} + V_{ef} = v$$
$$I_{bc} + I_{be} = I_{ab}$$

Note: In a circuit with N nodes, the number of branches B will always be $(N-1)+L$, where, L are maximum number of independent closed paths possible in the circuit.

Hence, we can always form B equations using Ohm's law, $(N-1)$ equations using KCL, L equations using KVL. Hence in total $2B$ equations can be formed, which are sufficient to solve for $2B$ variables (voltage and current in each branch).

Can we simplify the situation? Loop currents method: We do away with branch currents and define loop currents. The branch currents can be written in terms of loop currents once all the loop currents passing through the branch and their directions are known. The branch voltages can always be written using Ohm's law and branch current written in terms of loop currents. So now our objective is to find loop currents. For this we choose maximum number of independent loops and apply KVL in them.

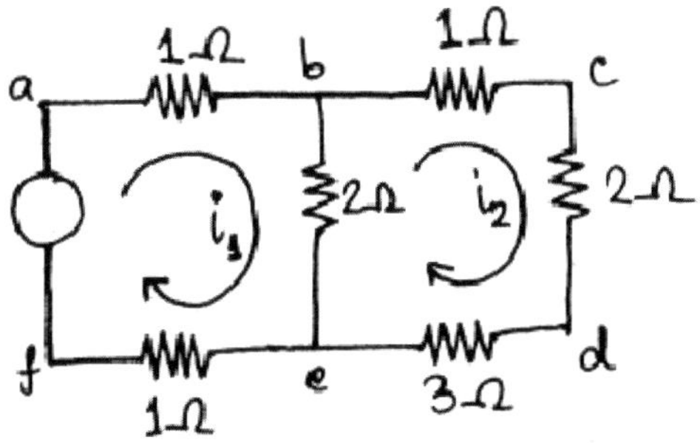

If i_1 and i_2 are known, voltages across all elements can be found.

Make two independent equations:

For loop abef:

$$v - i_1 \times 1 - (i_1 - i_2) \times 2 - i_1 \times 1 = 0$$

$$v = i_1 (1 + 2 + 1) - 2i_2$$

For loop bcde:

$$-(i_2 - i_1)2 - 1i_2 - 2i_2 - 3i_3 = 0$$

$$0 = -2i_2 + i_2 (2 + 1 + 2 + 3)$$

Use any technique to solve these (such as using matrices). We get:

$$i_1 = \frac{2}{7} Amperes$$

$$i_2 = \frac{2}{7} Amperes$$

Nodal Voltage Method

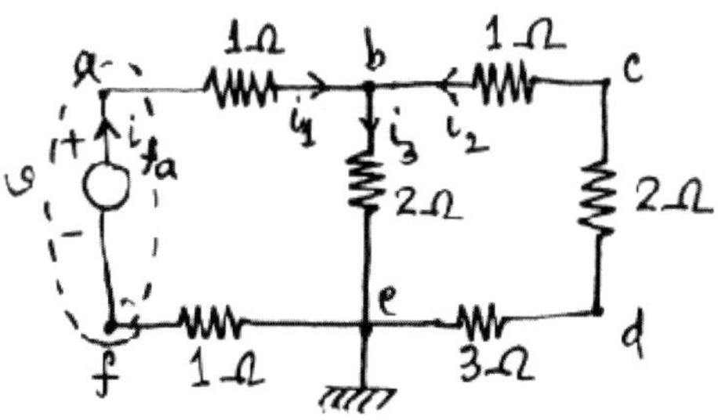

Nodal Voltage method

Independent Nodes: One of the nodes in circuit need to be considered as reference node. Hence its node potential is zero. For other nodes, nodal voltage is potential difſeretial w.r.t. to reference node. The nodes are called independent nodes. In general for N node network, $N-1$ nodes will be independent.

At node b: $i_1 + i_2 - i_3 = 0$

Similarly, other equations are:

$$\frac{E_a - E_b}{1} - \frac{E_c - E_b}{1} - \frac{E_b}{2} = 0$$

$$\frac{E_a - E_b}{1} = i_{fa}$$

$$E_a = E_f + v$$

$$\frac{-E_f}{1} = i_{fa}$$

$$\frac{E_c - E_b}{1} = \frac{E_d - E_c}{2}$$

$$\frac{-E_d}{3} = \frac{E_d - E_c}{2}$$

These are above six equations, in six unknowns. Thus can be solved for a unique solution. One can make a supernode and use KCL combining the nodes nodes a and f. We also make extra equations for potential difference between two nodes. With supernode, no. of equations is equal to no. of independent nodes whose voltage w.r.t. reference needs to be determined.

Current Sources in Loop Current Analysis

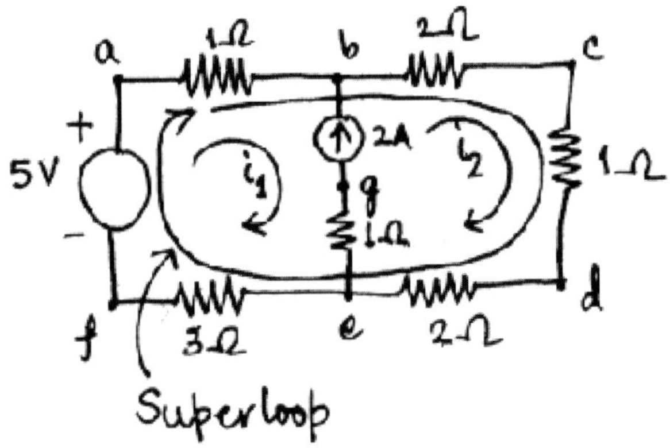

Figure: Current sources in Loop current analysis

Using KVL for loop 1 in figure:

$$5-i_1-v_{bg}-(i_1-i_2)1-i_1 3=0$$

The other equations are:

$$i_1-i_2=2$$

$$v_{bg}-i_2 2-i_2 1-i_2 2-(i_2-i_1)=0$$

The first and the third equation can be combined for taking care of v_{bg}. This can be done by making superloop for writing KVL.

Graph

For analysing circuits efficiently.

Loop Current Method

One can form a spanning tree from graph such that current sources are in links (Those elements which do not form part of the tree). Each link when added to the tree gives a loop. All voltage sources should be kept in branches of tree. For example, refer to the following two figures:

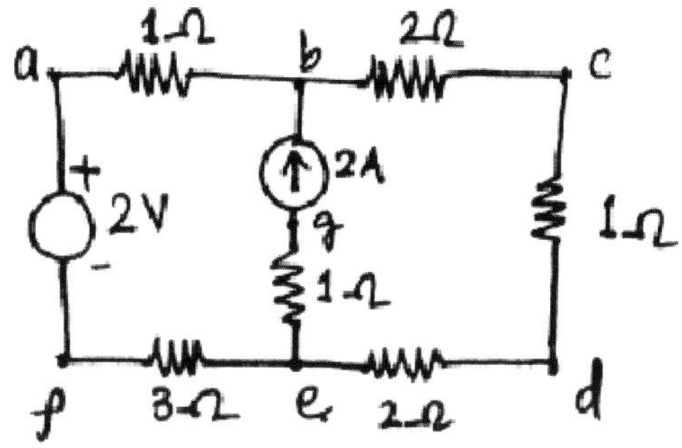

Figure: Loop Current method

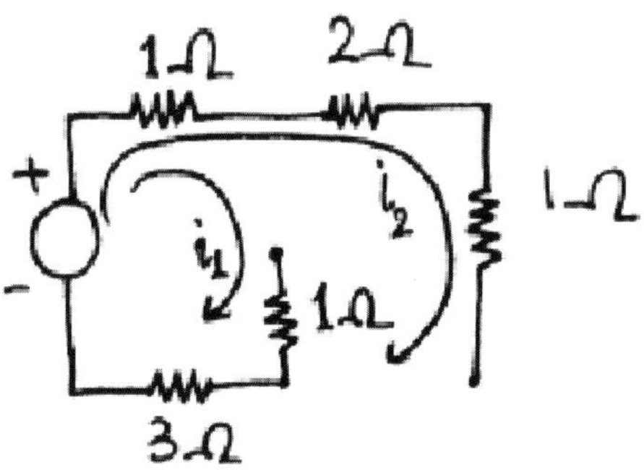

Figure: Loop current method

Node Voltages

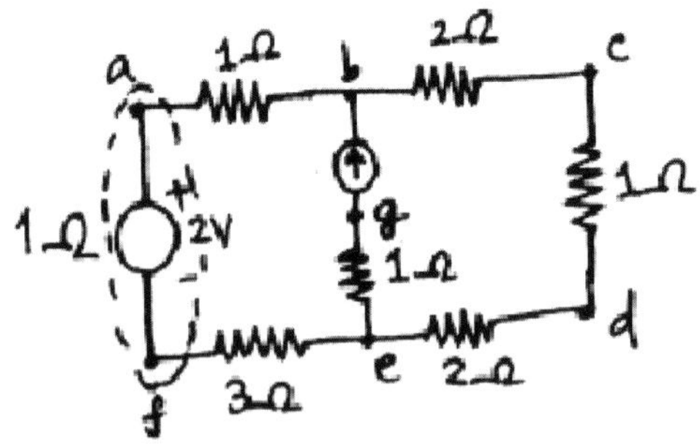

Figure: Node Voltage method

In the above figure, $V_{af} = 2V$. There are five unknown node voltages in the above circuit, namely, V_{ab}, V_{fe}, V_{bc}, V_{cd} and V_{ge}. Correspondingly, we have five equations. Note that we can merge af into one supernode

$$V_{ab} + V_{fe}/3 = 0$$
$$V_{ab} - V_{bc}/2 + 2 = 0$$
$$V_{bc}/2 - V_{cd} = 0$$
$$V_{cd} - V_{de}/2 = 0$$
$$V_{ge} + 2 = 0$$

The second equation follows from looking at node b, while the third one from doing the same at node c.

AC Circuit

The path for the flow of alternating current is called an AC Circuit. The alternating current (AC) is used for domestic and industrial purposes. In an AC circuit, the value of the magnitude and the direction of current and voltages is not constant, it changes at a regular interval of time. It travels as a sinusoidal wave completing one cycle as half positive and half negative cycle and is a function of time (t) or angle (θ=wt).

In DC Circuit, the opposition to the flow of current is the only resistance of the circuit whereas the opposition to the flow of current in the AC circuit is because of resistance (R), Inductive Reactance (X_L=2πfL) and capacitive reactance (X_C = 1/2 πfC) of the circuit.

In AC Circuit, the current and voltages are represented by magnitude and direction. The alternating quantity may or may not be in phase with each other depending upon the various parameters of the circuit like resistance, inductance, and capacitance. The sinusoidal alternating quantities are voltage and current which varies according to the sine of angle θ.

For the generation of electric power, in all over the world the sinusoidal voltage and current are selected because of the following reasons are given below:

- The sinusoidal voltage and current produce low iron and copper losses in the transformer and rotating electrical machines, which in turns improves the efficiency of the AC machines.
- They offer less interference to the nearby communication system.
- They produce less disturbance in the electrical circuit.

Alternating Voltage and Current in an AC Circuit

The voltage that changes its polarity and magnitude at regular interval of time is called an alternating voltage. Similarly the direction of the current is changed and the magnitude of current changes with time it is called alternating current.When an alternating voltage source is connected across a load resistance as shown in the figure below, the current through it flows in one direction and then in the opposite direction when the polarity is reversed.

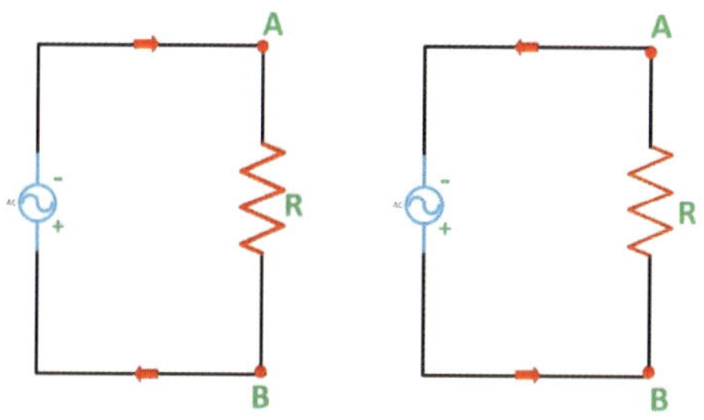

Alternating Currnt Circuit Diagram

The waveform of the alternating voltage with respect to the time and the current flowing through the resistance (R) in the circuit is shown below.

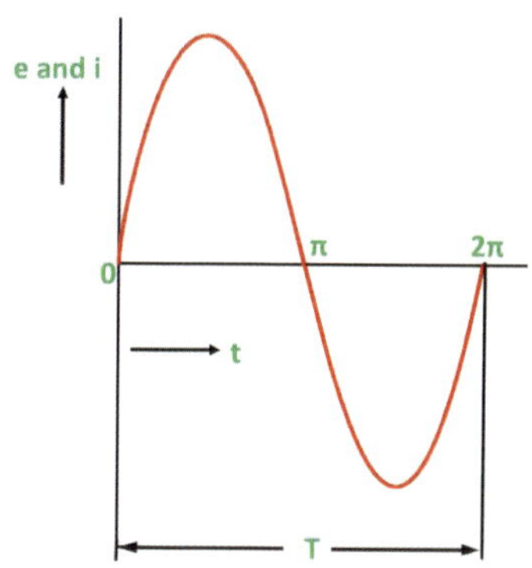

There are various types of AC circuit such as AC circuit containing only resistance (R), AC circuit containing only capacitance (C), AC circuit containing only inductance (L), the combination of RL Circuit, AC circuit containing resistance and capacitance (RC), AC circuit containing inductance and capacitance (LC) and resistance inductance and capacitance (RLC) AC circuit.

The various terms which are frequently used in an AC Circuit are as follows:

- Amplitude

The maximum positive or negative value attained by an alternating quantity in one complete cycle is called Amplitude or peak value or maximum value. The maximum value of voltage and current is represented by E_m or V_m and I_m respectively.

- Alternation

One half cycle is termed as alternation. An alternation span is of 180 degrees electrical.

- Cycle

When one set of positive and negative values completes by an alternating quantity or it goes through 360 degrees electrical, it is said to have one complete Cycle.

- Instantaneous Value

The value of voltage or current at any instant of time is called an instantaneous value. It is denoted by (i or e).

- Frequency

The number of cycles made per second by an alternating quantity is called frequency. It is measured in cycle per second (c/s) or hertz (Hz) and is denoted by (f).

- Time Period

The time taken in seconds by a voltage or a current to complete one cycle is called Time Period. It is denoted by (T).

- Wave Form

The shape obtained by plotting the instantaneous values of an alternating quantity such as voltage and current along the y axis and the time (t) or angle (θ=wt) along the x axis is called waveform.

AC Circuit Analysis

The topological analysis of basic ac electric circuits containing impedances and ideal ac supplies are presented in the following subsections. As will be demonstrated, using phasors greatly simplifies the analysis, and the VIs provide a flexible self-learning tool allowing users to create different circuit scenarios.

Equivalent Impedances and Circuits

An ac circuit may contain a number of series and parallel branches. As will be studied in the following paragraphs, however, it is possible to divide any complex ac circuit into subcircuits that include simple circuit combinations. Therefore, five ac circuit combinations are identified and studied here: equivalent impedance circuit, voltage divider circuit, current divider circuit, series/parallel (combination) circuit, and circuit with dual ac supply.

The equivalent impedance of any number of impedances in series or in parallel is the sum of the individual impedances or the sum of the admittances that is equal to $1/Z$, respectively.

Equation:

$$Z_{eq(series)} = Z_1 + Z_2 + Z_3 + Z_4$$

Equation:

$$Z_{eq(parallel)} \frac{1}{\left(1/Z_1 + 1/Z_2 + 1/Z_3 + 1/Z_4\right)}$$

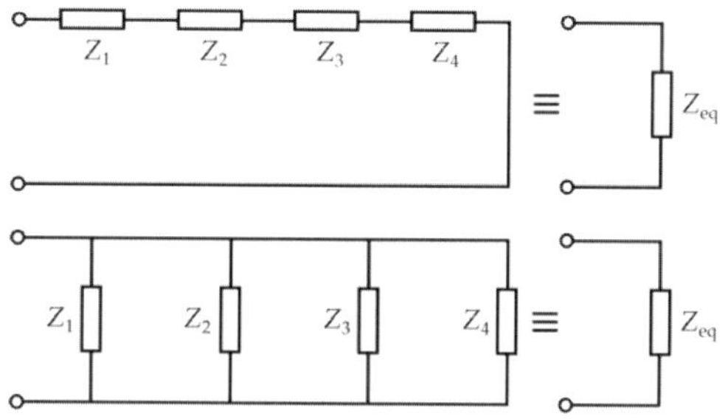

Figure: Series and parallel impedance circuits.

Since, the current phasors and voltage phasors are related by complex impedances, Kirchhoff's Voltage and Current Laws can be applied to ac circuits containing sinusoidal sources operating in steady-state.

The voltage divider circuit in figure (a) below indicates that if two impedances are connected in series and share the same current (meaning no other element is connected to the node where Z1 and Z2 join), the voltages across each of the elements are proportional to their impedances. The current and the voltages across each impedance can be given as:

Equation

$$I = \frac{V_s}{\left(Z_1 + Z_2\right)}$$

Equation

$$V_1 = Z_1 I$$

Equation

$$V_2 = Z_2 I$$

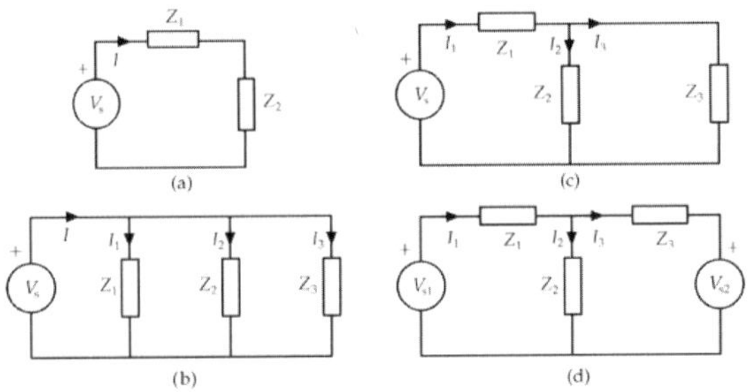

Figure: Circuits with impedances: (a) voltage divider, (b) current divider, (c) series/parallel circuit, and (d) circuit with dual supplies.

Let's consider the current divider circuit in now. Since the voltages across the impedances are equal to the supply voltage,

Equation

$$V_s = V_1 = V_2 = V_3$$

then the currents in each impedance branch can be calculated easily.

Equation

$$I_1 = \frac{V_s}{Z_1}, \quad I_2 = \frac{V_s}{Z_2}, \quad I_3 = \frac{V_s}{Z_3}$$

Hence, the supply current I is simply the sum of all three currents.

Equation

$$I = I_1 + I_2 + I_3 \quad or \quad I = \frac{V_s}{Z_{eq}}$$

The series/parallel circuit of above figure (c) can be analyzed easily if the equivalent impedance is calculated first.

Equation

$$Z_{eq} = Z_1 + \frac{Z_2 \bullet Z_3}{Z_2 + Z_3}$$

All remaining unknown parameters in the circuit can be derived as:

Equation

$$I_1 = \frac{V_s}{Z_{eq}}, \quad I_2 = \frac{V_2}{Z_2}, \quad I_3 = \frac{V_3}{Z_3}$$

Equation

$$V_1 = I_1 Z_1, \quad V_2 = V_3 = V_s - V_1, \quad V_3 = I_3 Z_3$$

The circuit given in above figure (d) is selected since it represents a commonly used equivalent circuit in electrical engineering: single-phase transformer, and asynchronous motor. Note that the circuit given in above figure (c) can be obtained easily by eliminating the supply voltage V_{s2} in above figure (d).

Hence the supply voltages are given as:

Equation

$$V_{s1} - I_1 (Z_1 + Z_2) - I_3 Z_2$$

Equation

$$V_{s2} = I_1 Z_2 - I_3 (Z_2 + Z_3)$$

The currents and voltages of the impedance elements can be determined after solving just above two equations, in which the known parameters are V_{s1}, V_{s2}, Z_1, Z_2, and Z_3, and the unknowns are I_1 and I_3. To solve for I_1 and I_3, the back substitution method is used in the VI provided in the following section. Then the remaining unknowns, I_2, V_1, V_2, and V_3, in the circuit are calculated as:

Equation

$$I_2 = I_1 - I_3$$

Equation

$$V_1 = Z_1 I_1, \quad V_2 = Z_2 I_2, \quad V_3 = Z_3 I_3$$

Virtual Instrument Panel

The virtual instruments provided in this topic contain six circuit options. The desired circuits can be selected from the available folders. When a VI is selected, the corresponding circuit and its associated controls are displayed with their default values. Then students can run the VI and vary the complex quantities of the circuit and observe the calculated values all in the complex forms.

As demonstrated, the parameters in the circuits are given in rectangular form, which can easily be related to the practical values. In addition, the rectangular form allows the user to omit either a real or an imaginary component of the complex numbers. For a sample front panel.

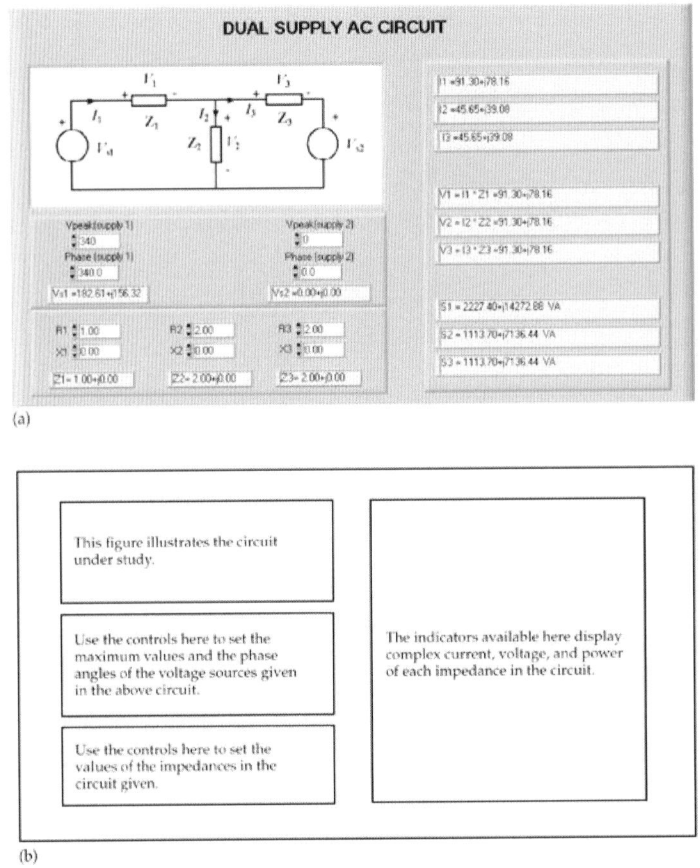

A sample front panel and brief user guide.

A Reverse Study

A reverse study is presented here, where an unknown impedance is determined utilizing ac voltage and current waveforms that are set by the user.

Illustrates the block diagram of the VI model, where the ideal sine wave current i_s and voltage v_s waveforms are defined by the user, and the corresponding equivalent impedance seen from the two terminals A–B is determined.

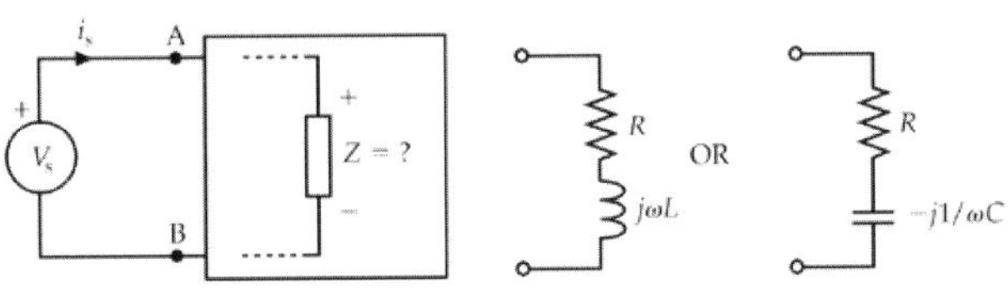

Figure: The circuit simulated to identify the equivalent impedance of an unknown load Z.

The complex value of the equivalent impedance is displayed on the front panel, which is used to interpret the nature of the load (R, L, C, R + L, or R + C). The phasor diagram is also provided in the VI to illustrate the voltage and impedance phasors.

Virtual Instrument Panel

Illustrates the front panel and the explanations diagram of the VI under investigation. The VI provided here (Single Phase AC Definitions.vi) uses the voltage and current as the base of the per-unit system, mainly for display purposes. The equivalent impedance is calculated and displayed on the front panel (as a per-unit value and a real value).

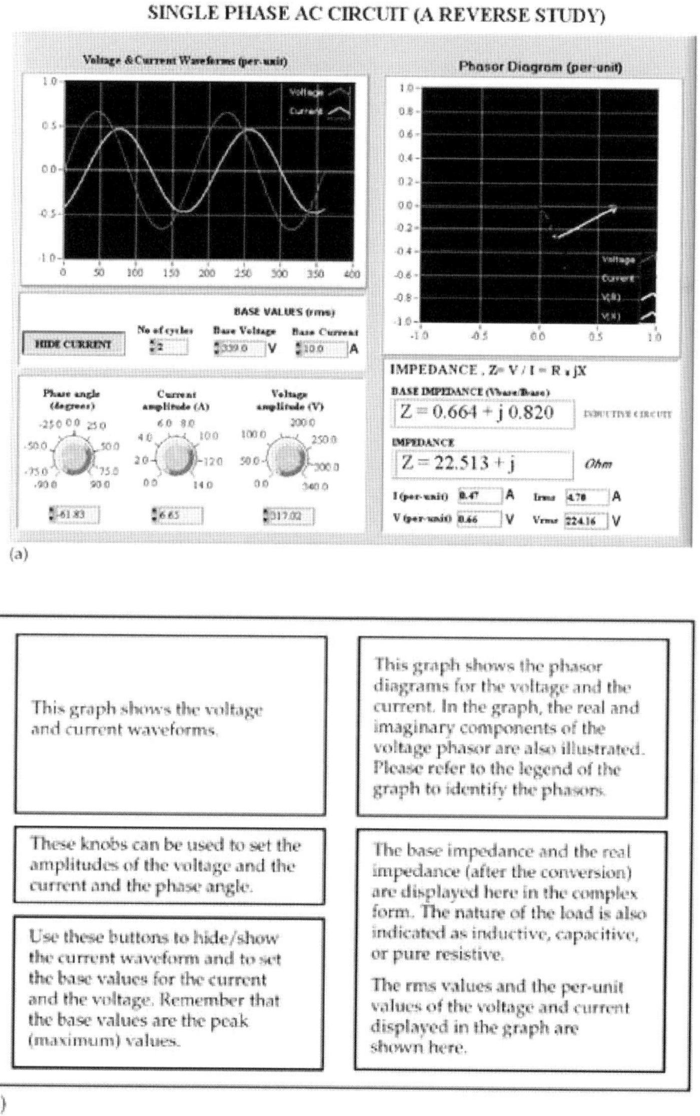

The front panel and the brief user guide of Single Phase AC Definitions.vi.

Resistance

Resistance is a measure of the opposition to current flow in an electrical circuit. Resistance is measured in ohms, symbolized by the Greek letter omega (Ω). Ohms are named after Georg Simon Ohm (1784-1854), a German physicist who studied the relationship between voltage, current and resistance. He is credited for formulating Ohm's Law.

Ohm's Law

The mathematical equation of Ohm's law may be written as:

$$R = \frac{V}{I}$$

where,

R is the resistance of the object, measured in Ohms, equivalent to $J \cdot s/C^2$

V is the potential difference across the object, measured in Volts

I is the current through the object, measured in Amperes

The ratio of voltage divided by electric current is also called the chordal resistance.

Resistance of a Conductor

DC Resistance

As long as the current density is totally uniform, the DC resistance R of a conductor of regular cross-section can be computed as:

$$R = \frac{\ell \cdot \rho}{A}$$

where,

ℓ is the length of the conductor, measured in meters

A is the cross-sectional area, measured in square meters

ρ is the electrical resistivity (also called *specific electrical resistance*) of the material, measured in Ohm • meter. Resistivity is a measure of the material's ability to oppose electric current.

For practical reasons, almost any connections to a real conductor will almost certainly mean the current density is not totally uniform. However, this formula still provides a good approximation for long thin conductors such as wires.

AC Resistance

If a wire conducts high-frequency alternating current then the effective cross sectional area of the wire is reduced because of the skin effect. This causes the wire resistance to increase at a rate of 10dB/decade for wire radius much greater than skin depth.

In a conductor close to others, the actual resistance is higher than that predicted by the skin effect because of the proximity effect.

Examples: Current must struggle to flow through the small coils of a toaster, enough to generate heat that browns bread. Old-style incandescent light bulbs force current to flow through filaments so thin that light is generated.

Resistance cannot be measured in an operating circuit. Accordingly, troubleshooting technicians often determine resistance by taking voltage and current measurements and applying Ohm's Law:

$$E = I \times R$$

That is, volts = amps x ohms). R stands for resistance in this formula. If resistance is unknown, the formula can be converted to R = E/I (ohms = volts divided by amps).

Examples: In an electric heater circuit, as portrayed in the two illustrations below, resistance is determined by measuring circuit voltage and current, then applying Ohm's Law.

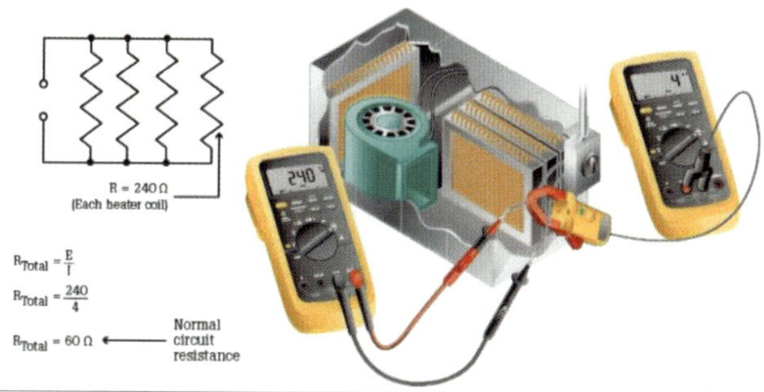

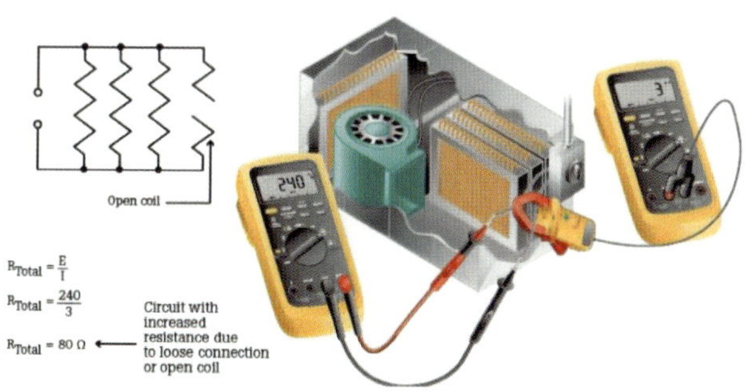

In the first example, total normal circuit resistance, a known reference value, is 60 Ω (240 ÷ 4 = 60 Ω). The 60 Ω resistance can help determine the condition of a circuit.

In the second example, if circuit current is 3 amps instead of 4, circuit resistance has increased from 60 Ω to 80 Ω (240 ÷ 3 = 80 Ω). The 20 Ω gain in total resistance could be caused by a loose or dirty connection or an open-coil section. Open-coil sections increase the total circuit resistance, which decreased current.

Causes of Resistance

In Metals

A metal consists of a lattice of atoms, each with a shell of electrons. This can also be known as a positive ionic lattice. The outer electrons are free to dissociate from their parent atoms and travel through the lattice, creating a 'sea' of electrons, making the metal a conductor. When an electrical potential difference (a voltage) is applied across the metal, the electrons drift from one end of the conductor to the other under the influence of the electric field.

Near room temperatures, the thermal motion of ions is the primary source of scattering of electrons (due to destructive interference of free electron wave on non-correlating potentials of ions), and is thus the prime cause of metal resistance. Imperfections of lattice also contribute into resistance, although their contribution in pure metals is negligible.

The larger the cross-sectional area of the conductor, the more electrons are available to carry the current, so the lower the resistance. The longer the conductor, the more scattering events occur in each electron's path through the material, so the higher the resistance. Different materials also affect the resistance.

In Semiconductors and Insulators

In metals, the Fermi level lies in the conduction band giving rise to free conduction electrons. However, in semiconductors the position of the Fermi level is within the band gap, approximately half-way between the conduction band minimum and valence band maximum for intrinsic (undoped) semiconductors. This means that at 0 Kelvin, there are no free conduction electrons and the resistance is infinite. However, the resistance will continue to decrease as the charge carrier density in the conduction band increases. In extrinsic (doped) semiconductors, dopant atoms increase the majority charge carrier concentration by donating electrons to the conduction band or accepting holes in the valence band. For both types of donor or acceptor atoms, increasing the dopant density leads to a reduction in the resistance. Highly doped semiconductors hence behave metallic. At very high temperatures, the contribution of thermally generated carriers will dominate over the contribution from dopant atoms and the resistance will decrease exponentially with temperature.

In Ionic Liquids/Electrolytes

In electrolytes, electrical conduction happens not by band electrons or holes, but by full atomic species (ions) traveling, each carrying an electrical charge. The resistivity of ionic liquids varies tremendously by the salt concentration—while distilled water is almost an insulator, salt water is a very efficient electrical conductor. In biological membranes, currents are carried by ionic salts. Small holes in the membranes, called ion channels, are selective to specific ions and determine the membrane resistance.

Resistance of Various Materials

Material	Resistivity, ρ ohm-meter
Metals	10^{-8}
Semiconductors	variable
Electrolytes	variable
Insulators	10^{16}
Superconductors	0 (exactly)

Band Theory Simplified

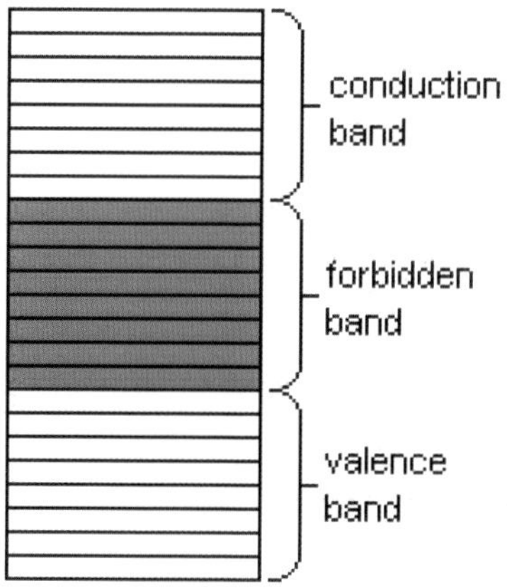

Electron energy levels in an insulator.

Quantum mechanics states that the energy of an electron in an atom cannot be any arbitrary value. Rather, there are fixed energy levels which the electrons can occupy, and values in between these levels are impossible. The energy levels are grouped into two bands: the valence band and the conduction band (the latter is generally above the former). Electrons in the conduction band may move freely throughout the substance in the presence of an electrical field.

In insulators and semiconductors, the atoms in the substance influence each other so that between the valence band and the conduction band there exists a forbidden band of energy levels, which the electrons cannot occupy. In order for a current to flow, a relatively large amount of energy must be furnished to an electron for it to leap across this forbidden gap and into the conduction band. Thus, even large voltages can yield relatively small currents.

Differential Resistance

When, resistance may depend on voltage and current, differential resistance, incremental resistance or slope resistance is defined as the slope of the *V-I* graph at a particular point, thus:

$$R = \frac{dV}{dI}$$

This quantity is sometimes called simply *resistance*, although the two definitions are equivalent only for an ohmic component such as an ideal resistor. For example, a diode is a circuit element for which the resistance depends on the applied voltage or current.

If the *V-I* graph is not monotonic (i.e. it has a peak or a trough), the differential resistance will be negative for some values of voltage and current. This property is often known as *negative resistance*, although it is more correctly called negative differential resistance, since the absolute resistance V/I is still positive. Example of such an element is a tunnel diode.

Temperature-dependence

Near room temperature, the electric resistance of a typical metal varies linearly with the temperature. At lower temperatures (less than the Debye temperature), the resistance decreases as T^5 due to the electrons scattering off of phonons. At even lower temperatures, the dominant scattering mechanism for electrons is other electrons, and the resistance decreases as T^2. At some point, the impurities in the metal will dominate the behavior of the electrical resistance which causes it to saturate to a constant value. Matthiessen's Rule (first formulated by Augustus Matthiessen in the 1860s; the equation below gives its modern form) says that all these different types of behavior can be summed up to get the total resistance as a function of temperature,

$$R = R_{imp} + aT^2 + bT^5 + cT$$

where, R_{imp} is the temperature independent electrical resistivity due to impurities, and a, b, and c are coefficients which depend upon the metal's properties. This rule can be seen as the motivation to Heike Kamerlingh Onnes's experiments that lead in 1911 to the discovery of superconductivity.

The electric resistance of a typical intrinsic (non doped) semiconductor decreases exponentially with the temperature:

$$R = R_0 e^{-aT}$$

Extrinsic (doped) semiconductors have a far more complicated temperature profile.

As temperature increases starting from absolute zero they first decrease steeply in resistance as the carriers leave the donors or acceptors. After most of the donors or acceptors have lost their carriers the resistance starts to increase again slightly due to the reducing mobility of carriers (much as in a metal). At higher temperatures it will behave like intrinsic semiconductors as the carriers from the donors/acceptors become insignificant compared to the thermally generated carriers.

The electric resistance of electrolytes and insulators is highly nonlinear, and case by case dependent, therefore no generalized equations are given.

Measuring Resistance

An instrument for measuring resistance is called a resistance meter or ohmmeter. Simple ohmmeters cannot measure low resistances accurately because the resistance of their measuring leads causes a voltage drop that interferes with the measurement, so more accurate devices use four-terminal sensing.

Impedance

The term impedance is common used if someone connects a loudspeaker (amplifier) to an audio system; it is normally a number of Ohms, regularly printed next to many inputs or to an output socket. Though the property of impedance is less understood, the word impedance is used in many engineering disciplines to refer to as an opponent to work done.Anyway, this article refers particularly to electrical impedance, which describes a combined effect of resistance (R), inductive reactance (XL) and capacitive reactance (XC) in an AC circuit, whether it is occurring in a single component, or in an entire circuit.

Electrical Impedance

Electrical impedance (also known as "impedance" in short) is an addition of the definition of resistance to analternating current(AC). Which means that impedance includes both resistance (opposition of the electric current that causes heat) and reactance (a measure of such an oppositional current alternates) – in detail, the opposition adjacent to the electric currents. In the direct current (DC), electrical impedance is same as a resistance, except that it does not hold true in AC circuits.

Impedance can also be dissimilar from resistance when a DC circuit changes flow in one way or other- similar to the opening and closing of an electrical switch, as is observed in the computers when they open and close switches to represent ones and zeros (binary language). The opposite of an impedance is admittance, which is the measure of the allowance of current. The figure to the left is a complex impedance plane, in which impedance is represented by a Z, resistance is depicted as R, and the reactance is depicted with X.

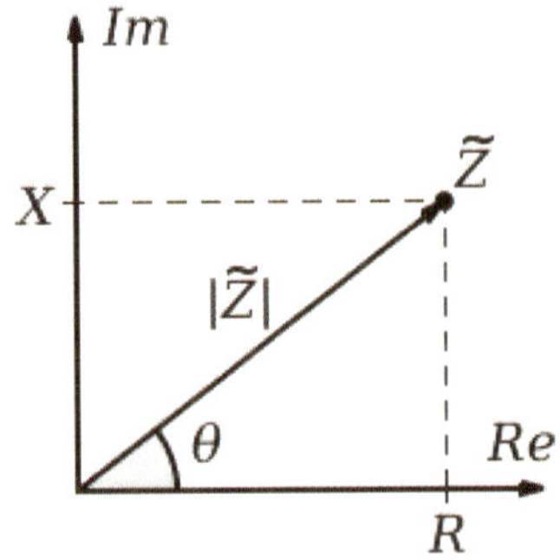

Electrical Impedance

Electrical Impedance Tomography (EIT)

The fundamental principle of electrical impedance tomography (EIT) is akin to electrical resistance tomography (ERT) such that several measurements at the periphery of a process vessel or tube are taken and combined to give information on the electrical properties of the process volume.

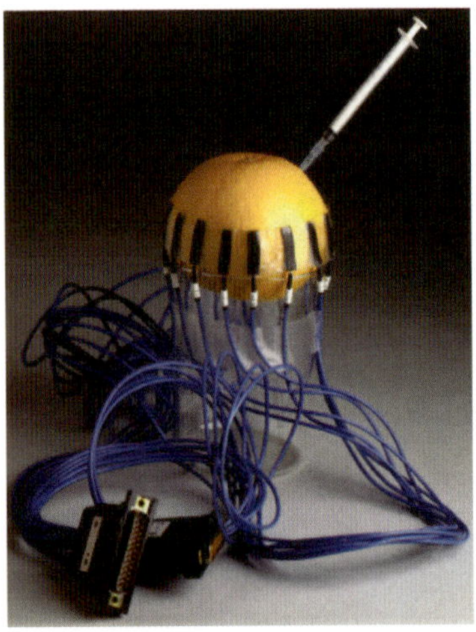

Electrical Impedance Tomography

Electrical Impedance Tomography (EIT) is a non-invasive medical imaging method in which a figure of the conductivity or permitivity of a part of the body is incidental from the surface electrode measurements. Electrical conductivity depends on the free ion content and differs significantly between different biological tissues (absolute EIT) or dissimilar practical states of one and other similar tissues or organs (relative or functional EIT). Majority of EIT systems apply little irregular currents at a single frequency; however, some EIT systems use various frequencies to better discriminate between usual and suspected abnormal tissue within the same organ (multifrequency-EIT or electrical impedance spectroscopy).

Complex Impedance

A resistor with a value of R has an impedance of R ohms, a real number. An ideal inductor has a complex impedance of:

$$Z=j2\pi fL$$

Where, 'f' is the frequency in Hertz and L is the inductance in Henries. It is imaginary because an a ideal inductor can simply store and release electrical energy. It can't dissipate it as heat like a resistor. Similarly, an ideal capacitor has a complex impedance of:

$$Z = -j/2\pi fc$$

Where, 'C' is the capacitance in farads.

Use of Complex Impedance

The behavior of the impedance of an AC circuit with various components quickly becomes unmanageable if sines and cosines are used to present the voltages and current. A mathematical build which eases the complexity use of complex exponential functions. The necessary parts of the strategy are as follows:

Math relationship underlaying the technique

$$ej\omega t = \cos\omega t + \sin\omega t$$

The real part of a complex exponential function can be used to represent an AC voltage or current.

$$V = Vm\ COS\omega t$$

$$I = Im\ COS\ (\omega t\text{-}\varphi)$$

The impedance can then be expressed as a complex exponential:

$$Z = Vm/\ Im\ e\text{-}j\emptyset = R+jX$$

The impedance of the individual circuit elements can then be expressed as pure real or imaginary numbers:

$$R\ -j/\omega c\ j\omega L$$

Complex Impedance for RL and RC

Using complex impedance is a significant technique for handling multi-component AC circuits. If a complex plane is used with resistance along the real axis, then the reactance of a capacitor and inductor are treated as imaginary numbers. For series combinations of the components such as RL and RC combinations, the component values are added as if they were components of a vector. Shown now is the Cartesian form of the complex impedance. They can also be written in polar form. Impedances in combination circuits like the RLC parallel circuit.

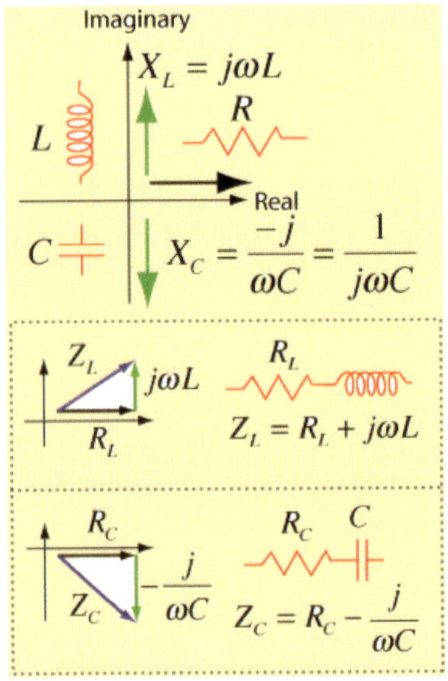

Applications of Impedance

Impedance and resistance both have applications whether you consider it or not, both exist in your own house. Your house electricity is controlled by a panel that has fuses in it. When you go through an electrical surge, the fuses are there to interrupt the power so that the injury is minimized. Your fuses are similar to very high-capacity resistors that are capable of taking the blow. Without them, your house's electrical system would fry and you would have to make it up from scratch.

This problem can be solved thanks to impedance and resistance. Another situation in which impedance has importance is in capacitors. In capacitors, impedance is used to manage the flow of electricity in a circuit board. Without the capacitors controlling and adaptable electrical flow, your electronics that use alternating currents will either fry or go berserk. Since alternating current delivers electricity at a fluctuating pulse, there needs to be a gate that holds back all the electricity and lets it go smoothly so that the electrical circuit is not overloaded or under loaded.

Transfer Functions

A control system consists of an output as well as an input signal. The output is related to the input through a function call transfer function. This function is represented by a block and the complete diagram of control system using these blocks which represent transfer function and arrows which represent various signals, is collectively known as block diagram of a control system. For any control system there exists a reference input termed as excitation or cause which operates through a transfer operation termed as transfer function and produces an effect resulting in controlled output or response. Thus the cause and effect relationship between the output and input is related to each other through a transfer function.

In Laplace Transform, if the input is represented by R(s) and output is represented by C(s), then the transfer function will be:

$$G(s) = \frac{C(s)}{R(s)} \Rightarrow R(s).G(s) = C(s)$$

That is, transfer function of the system multiplied by input function gives the output function of the system.

Input, Output and other Signals are Represented in Laplace form in a Control System

It is not necessary that output and input of a control system are of same category. For example, in electric motors the input is electrical signal whereas the output is mechanical signal since electrical energy required to rotate the motors. Similarly in an electric generator, the input is mechanical signal and the output is electrical signal, since mechanical energy is required to produce electricity in a generator.

But for mathematical analysis, of a system all kinds of signals should be represented in a similar form. This is done by transforming all kinds of signal to their Laplace form. Also the transfer function of a system is represented by Laplace form by dividing output Laplace transfer function to input Laplace transfer function. Hence a basic block diagram of a control system can be represented as:

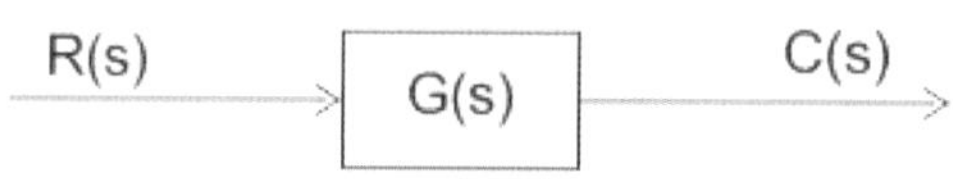

$$Where,\ R(s) = \pounds r(t),\ C(s) = \pounds c(t) \,\&\, G(s) = \frac{\pounds c(t)}{\pounds r(t)}$$

Where, r(t) and c(t) are time domain function of input and output signal respectively.

Transfer Function

The transfer function of a control system is defined as the ration of the Laplace transform of the output variable to Laplace transform of the input variable assuming all initial conditions to be zero.

$$G(s)=\frac{C(s)}{R(s)}$$

Procedure for determining the transfer function of a control system are as follows:

- We form the equations for the system.
- Now we take Laplace transform of the system equations, assuming initial conditions as zero.
- Specify system output and input.
- Lastly we take the ratio of the Laplace transform of the output and the Laplace transform of the input which is the required transfer function.

Methods of Obtaining a Transfer Function

There are major two ways of obtaining a transfer function for the control system. The ways are:

- Block Diagram Method : It is not convenient to derive a complete transfer function for a complex control system. Therefore the transfer function of each element of a control system is represented by a block diagram. Block diagram reduction techniques are applied to obtain the desired transfer function.
- Signal Flow Graphs : The modified form of a block diagram is a signal flow graph. Block diagram gives a pictorial representation of a control system. Signal flow graph further shortens the representation of a control system.

Poles and Zeros of Transfer Function

Generally a function can be represented to its polynomial form.

For example:

$$F(s)=f0s^{n}+f1s^{n-1}+f2s^{n-2}+f3s^{n-3}+......+f_{n-1}s^{1}+f_{n}$$

Now similarly transfer function of a control system can also be represented as:

$$G(s)=\frac{C(s)}{R(s)}=\frac{C_{0}s^{n}+C_{1}s^{n-1}+C_{2}s^{n-2}+.....+C_{n-1}s+C_{n}}{R_{0}s^{m}+R_{1}s^{m-1}+R_{2}s^{m-2}+...+R_{m}-1s+R_{m}}$$

$$=K\frac{(s-z_{1})(s-z_{2})(s-z_{3})...(s-z_{n})}{(s-p_{1})(s-p_{2})(s-p_{3})...(s-p_{n})}$$

Where, K is known as gain factor of the transfer function.

Now in the above function if s = z_1, or s = z_2, or s = z_3,....s = z_n, the value of transfer function becomes zero. These z_1, z_2, z_3,....z_n, are roots of the numerator polynomial. As for these roots the numerator polynomial, the transfer function becomes zero, these roots are called zeros of the transfer function. Now, if $s = p_1$, or $s = p_2$, or $s = p_3$,....$s = p_m$, the value of transfer function becomes infinite. Thus the roots of denominator are called the poles of the function. Now let us rewrite the transfer function in its polynomial form:

$$G(s) = K\frac{(s-z_1)(s-z_2)(s-z_3)...(s-z_n)}{(s-p_1)(s-p_2)(s-p_3)...(s-p_m)}$$

Now, let us consider s approaches to infinity as the roots are all finite number, they can be ignored compared to the infinite s. Therefore

$$G(s) = K\frac{s^n}{s^m} = Ks^{n-m}$$

Hence, when $s \to \infty$ and $n > m$, the function will have also value of infinity, that means the transfer function has poles at infinite s, and the multiplicity or order of such pole is n - m. Again, when $s \to \infty$ and $n < m$, the transfer function will have value of zero that means the transfer function has zeros at infinite s, and the multiplicity or order of such zeros is m - n.

Concept of Transfer Function

The transfer function is generally expressed in Laplace Transform and it is nothing but the relation between input and output of a system. Let us consider a system consists of a series connected resistance (R) and inductance (L) across a voltage source (V).

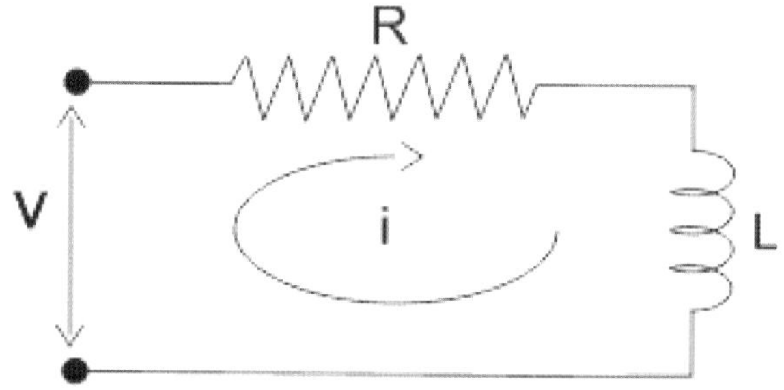

In this circuit, the current 'i' is the response due to applied voltage (V) as cause. Hence the voltage and current of the circuit can be considered as input and output of the system respectively. From the circuit, we get,

$$V = Ri + L\frac{di}{dt}$$

Now applying Laplace Transform, we get,

$$V(s) = RI(s) + L\left[sI(s) - i(0^+)\right]$$

$$\left[\because \textit{Initially inductor behaves as open, hence}, i(0^+) = 0\right]$$

$$\Rightarrow V(s) = I(s)[R + Ls]$$

$$\Rightarrow \frac{I(s)}{V(s)} = \frac{1}{R + Ls} = \frac{1/L}{s + R/L}$$

The transfer function of the system, G(s) = I(s)/V(s), the ratio of output to input:

- Let us explain the concept of poles and zeros of transfer function through an example.

$$G(s) = \frac{(s+1)(s+2)}{(s+3)(s+4)(s+5)(s+2-4j)(s+2+4j)}$$

Answer The zeros of the function are, -1, -2 and the poles of the functions are -3, -4, -5, -2 + 4j, -2 - 4j. Here n = 2 and m = 5, as n < m and m - n = 3, the function will have 3 zeros at $s \rightarrow \infty$. The poles and zeros are plotted in the figure below:

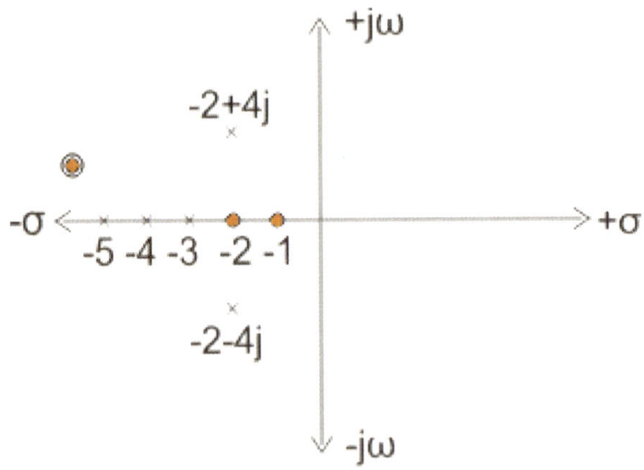

- Let us take another example of transfer function of control system

$$G(s) = \frac{(s-2)(s+5)(s+8)}{s(s+1)(s+6)(s+9)(s+1-j3)(s+1+j3)}$$

Answer: In the above transfer function, if the value of numerator is zero, then:

$$(s-2)(s+5)(s+8) = 0$$

$$\Rightarrow s = 2, -5, -8$$

These are the location of zeros of the function. Similarly, in the above transfer function, if the value of denominator is zero, then:

$$s(s+1)(s+6)(s+9)(s+1-j3)(s+1+j3)=0$$
$$\Rightarrow s=0,-1,-6,-9,-1+j3,-1-j3$$

These are the location of poles of the function.

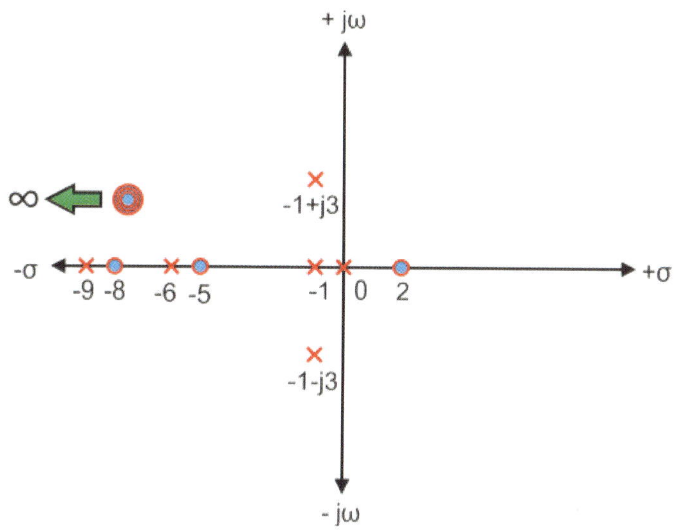

As the number of zeros should be equal to number of poles, the remaining three zeros are located at s →∞.

Example of Transfer Function of a Network

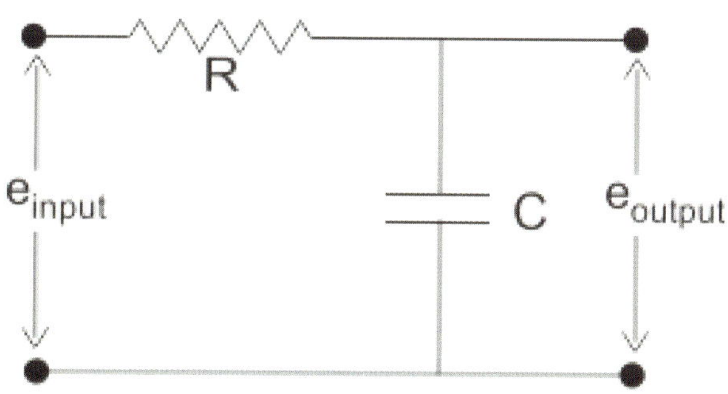

Answer: In the above network it is obvious that

$$e_{input} = Ri(t) + \frac{1}{C}\int i(t)dt$$

$$e_{output} = \frac{1}{C}\int i(t)dt$$

Let us assume,

$$£\left[e_{input}(t)\right]=E_{input}(s),£\left[e_{output}(t)\right]=E_{output}(s),£\left[i(t)\right]=I(s)$$

$$And\ then, £\left[\int i(t)dt\right]=\frac{I(s)}{s}+\int i(0)dt=\frac{I(s)}{s}+0=\frac{F(s)}{s}$$

$$\left[\int i(0)dt=0, as\ there\ is\ no\ current\ initially\ through\ the\ capacitor\right]$$

Taking the Laplace transform of above equations with considering the initial condition as zero, we get,

$$E_{input}(s)=RI(s)+\frac{1}{C}.\frac{I(s)}{s}=I(s)\left[R+\frac{1}{C_s}\right]$$

$$E_{output}(s)=\frac{1}{C}.\frac{I(s)}{s}=I(s)\frac{1}{C^s}$$

Transfer function of the network,

$$\frac{E_{output}(s)}{E_{input}(s)}=\frac{I(s)\frac{1}{C^s}}{I(s)\left[R+\frac{1}{C^s}\right]}=\frac{\frac{1}{C^s}}{\left[R+\frac{1}{C^s}\right]}=\frac{1}{RC_s+1}$$

The Effect of Impulse Signal

The unit impulse signal is defined as,

$$\delta(t)=1\ when\ t=0$$
$$=0\ when\ t\neq 0$$

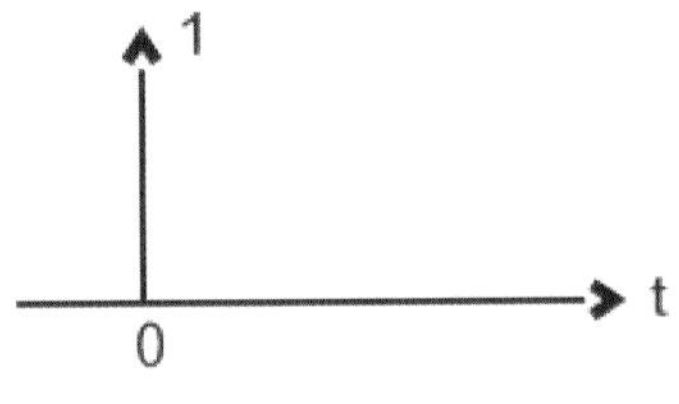

Laplace transform of unit impulse function is 1.

$$£\delta(t)=1$$

Now if input signal is unit impulse signal then,

$$\frac{C(s)}{R(s)}=G(s)\Rightarrow C(s)=R(s)G(s)\Rightarrow C(s)=G(s)$$

$$\left[\because £\delta(t)=1\right]$$

The output function is same as its transfer function.

Phasors

A *phasor* is a complex number in polar form that you can apply to circuit analysis. When you plot the amplitude and phase shift of a sinusoid in a complex plane, you form a phase vector, or phasor.

You can also think of a phasor as a rotating vector. Unlike a vector having magnitude and direction, a phasor has *magnitude* V_A and *angular displacement* ϕ. You measure angular displacement in the counterclockwise direction from the positive x-axis.

Here is a diagram of a voltage phasor as a rotating vector at some frequency, with its tail at the origin. If you need to add or subtract phasors, you can convert the vector into its *x*-component ($V_A \cos \phi$) and its *y*-component ($V_A \sin \phi$) with some trigonometry.

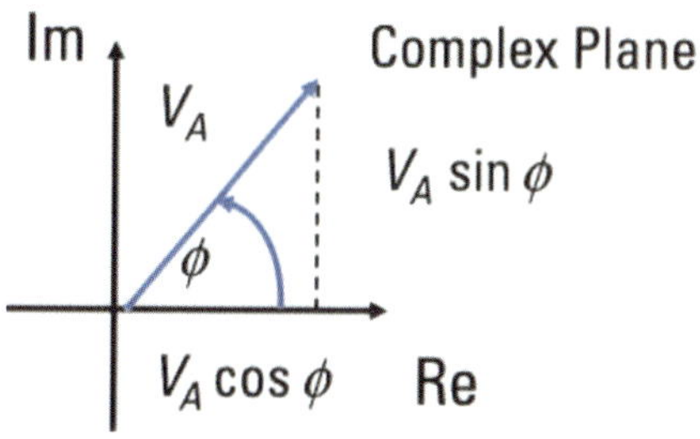

The following sections explain how to find the different forms of phasors and introduce you to the properties of phasors.

Find Phasor Forms

Phasors, which you describe with complex numbers, embody the amplitude and phase of a sinusoidal voltage or current. The phase is the angular shift of the sinusoid, which corresponds to a time shift t_o. So if you have $\cos[\omega(t - t_o)]$, then $\omega t_o = \ _o$, where ϕ_o is the angular phase shift.

To establish a connection between complex numbers and sine and cosine waves, you need the complex exponential $e^{j\theta}$ and Euler's formula:

$$e^{j\theta} = \cos\theta + j\sin\theta$$

where

$$j = \sqrt{-1}$$

The left side of Euler's formula is the *polar* phasor form, and the right side is the rectangular phasor form. You can write the cosine and sine as follows:

$$\cos\theta = \mathrm{Re}[e^{j\theta}]$$

$$\sin\theta = \mathrm{Im}[e^{j\theta}]$$

Here is a cosine function and a shifted cosine function with a phase shift of $\pi/2$.

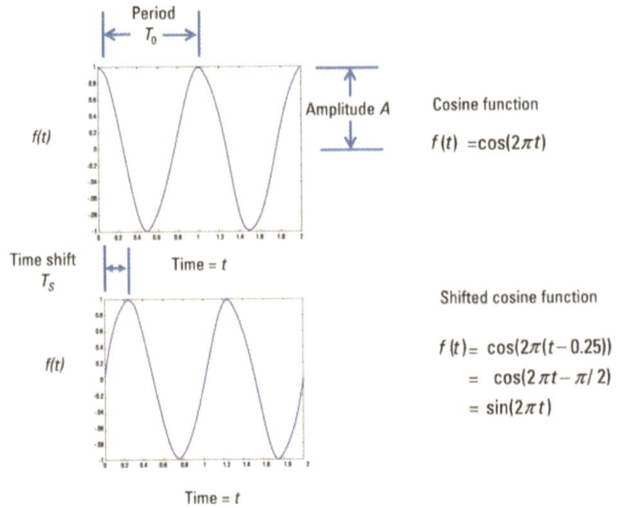

In general, for the sinusoids shown here, you have an amplitude V_A, a radian frequency ω, and a phase shift of ϕ given by the following expression:

$$v(t) = V_A \cos(\omega t + \phi)$$

$$v(t) = V_A \mathrm{Re}\left(e^{j(\omega t+\phi)}\right) = \mathrm{Re}\left[\underbrace{V_A e^{j\phi}} e^{j\omega t}\right]$$

Because the radian frequency ω remains the same in a linear circuit, a phasor just needs the amplitude V_A and the phase ϕ to get into polar form:

$$\mathrm{V} = V_A e^{j}\phi$$

To describe a phasor, you need only the amplitude and phase shift (not the radian frequency). Using Euler's formula, the rectangular form of the phasor is:

$$\mathrm{V} = V_A \cos\phi + jV_A \sin\phi$$

As the single vector rotates in an anti-clockwise direction, its tip at point A will rotate one complete revolution of 360° or 2π representing one complete cycle. If the length of its moving tip is transferred at different angular intervals in time to a graph as shown above, a sinusoidal waveform would be drawn starting at the left with zero time. Each position along the horizontal axis indicates the time that has elapsed since zero time, t = 0. When the vector is horizontal the tip of the vector represents the angles at 0°, 180° and at 360°.

Likewise, when the tip of the vector is vertical it represents the positive peak value, (+Am) at 90° or $\pi/2$ and the negative peak value, (-Am) at 270° or $3\pi/2$. Then the time axis of the waveform represents the angle either in degrees or radians through which the phasor has moved. So we can say that a phasor represent a scaled voltage or current value of a rotating vector which is "frozen" at some point in time, (t) and in our example above, this is at an angle of 30°.

Sometimes, when we are analysing alternating waveforms we may need to know the position of the phasor, representing the Alternating Quantity at some particular instant in time especially when we want to compare two different waveforms on the same axis. For example, voltage and current. We have assumed in the waveform above that the waveform starts at time t = 0 with a corresponding phase angle in either degrees or radians.

But if a second waveform starts to the left or to the right of this zero point or we want to represent in phasor notation the relationship between the two waveforms then we will need to take into account this phase difference, Φ of the waveform.

Phase Difference of a Sinusoidal Waveform

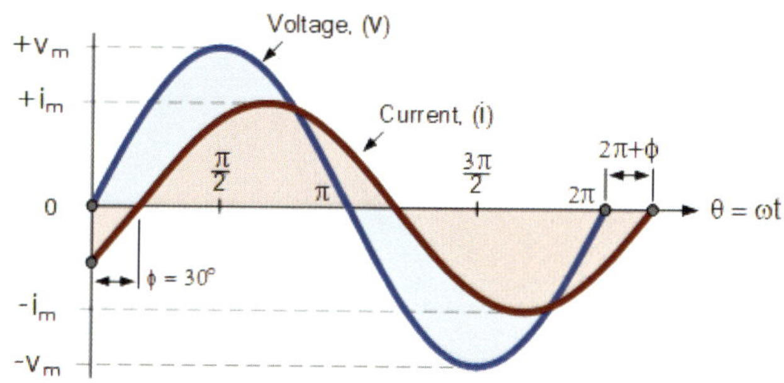

The generalised mathematical expression to define these two sinusoidal quantities will be written as:

$$V_{(t)} = V_m \sin(\omega t)$$
$$i_{(t)} = I_m \sin(\omega t - \phi)$$

The current, i is lagging the voltage, v by angle Φ and in our example above this is 30°. So the difference between the two phasors representing the two sinusoidal quantities is angle Φ.

Phasor Diagram of a Sinusoidal Waveform

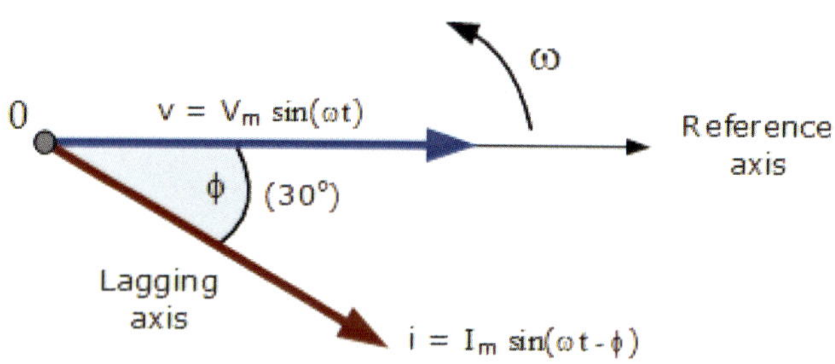

The phasor diagram is drawn corresponding to time zero (t = 0) on the horizontal axis. The lengths of the phasors are proportional to the values of the voltage, (V) and the current, (I) at the instant in time that the phasor diagram is drawn. The current phasor lags the voltage phasor by the angle, Φ, as the two phasors rotate in an *anti-clockwise*direction as stated earlier, therefore the angle, Φ is also measured in the same anticlockwise direction.

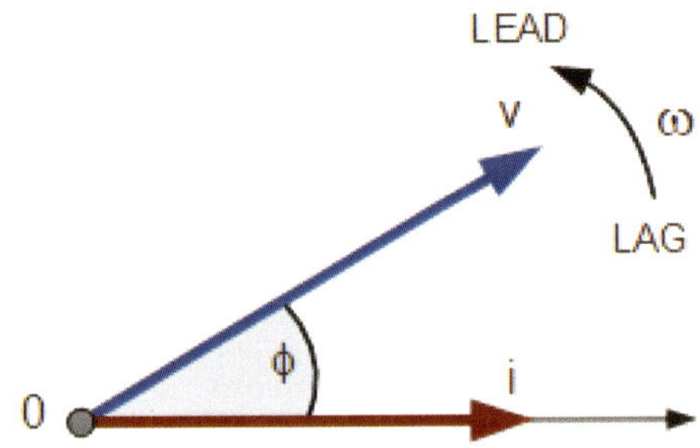

If however, the waveforms are frozen at time, t = 30°, the corresponding phasor diagram would look like the one shown on the right. Once again the current phasor lags behind the voltage phasor as the two waveforms are of the same frequency.

However, as the current waveform is now crossing the horizontal zero axis line at this instant in time we can use the current phasor as our new reference and correctly say that the voltage phasor is "leading" the current phasor by angle, Φ. Either way, one phasor is designated as the *reference* phasor and all the other phasors will be either leading or lagging with respect to this reference.

Phasor Addition

Sometimes, it is necessary when studying sinusoids to add together two alternating waveforms, for example in an AC series circuit, that are not in-phase with each other. If they are in-phase that is, there is no phase shift then they can be added together in the same way as DC values to find the algebraic sum of the two vectors. For example, if two voltages of say 50 volts and 25 volts respectively are together "in-phase", they will add or sum together to form one voltage of 75 volts (50 + 25).

If however, they are not in-phase that is, they do not have identical directions or starting point then the phase angle between them needs to be taken into account so they are added together using phasor diagrams to determine their Resultant Phasor or Vector Sum by using the *parallelogram law*.

Consider, two AC voltages, V_1 having a peak voltage of 20 volts, and V_2 having a peak voltage of 30 volts where V_1 leads V_2 by 60°. The total voltage, V_T of the two voltages can be found by firstly drawing a phasor diagram representing the two vectors and then constructing a parallelogram in which two of the sides are the voltages, V_1 and V_2 as shown below.

Phasor Addition of two Phasors

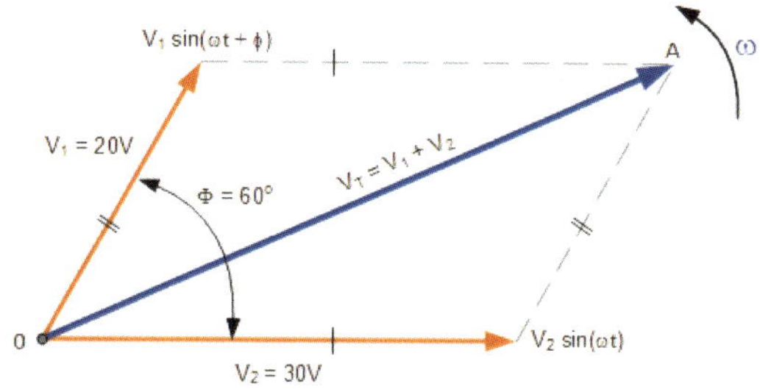

By drawing out the two phasors to scale onto graph paper, their phasor sum $V_1 + V_2$ can be easily found by measuring the length of the diagonal line, known as the "resultant r-vector", from the zero point to the intersection of the construction lines 0-A. The downside of this graphical method is that it is time consuming when drawing the phasors to scale.

Also, while this graphical method gives an answer which is accurate enough for most purposes, it may produce an error if not drawn accurately or correctly to scale. Then one way to ensure that the correct answer is always obtained is by an analytical method.

Mathematically we can add the two voltages together by firstly finding their "vertical" and "horizontal" directions, and from this we can then calculate both the "vertical" and "horizontal" components for the resultant "r vector", V_T. This analytical method which uses the cosine and sine rule to find this resultant value is commonly called the Rectangular form.

In the rectangular form, the phasor is divided up into a real part, x and an imaginary part, yforming the generalised expression Z = x ± jy. This then gives us a mathematical expression that represents both the magnitude and the phase of the sinusoidal voltage as:

Complex Sinusoid

$$V_m = \cos(\phi) + jV_m(\sin\phi)$$

So the addition of two vectors, A and B using the previous generalised expression is as follows:

$$A = x + jy \qquad B = w + jz$$
$$A + B = (x + w) + j(y + z)$$

Phasor Addition using Rectangular Form

Voltage, V_2 of 30 volts points in the reference direction along the horizontal zero axis, then it has a horizontal component but no vertical component as follows:

- Horizontal Component = 30 cos 0° = 30 volts

- Vertical Component = 30 sin 0° = 0 volts

This then gives us the rectangular expression for voltage V_2 of: 30 + j0

Voltage, V_1 of 20 volts leads voltage, V_2 by 60°, then it has both horizontal and vertical components as follows:

- Horizontal Component = 20 cos 60° = 20 x 0.5 = 10 volts
- Vertical Component = 20 sin 60° = 20 x 0.866 = 17.32 volts
- This then gives us the rectangular expression for voltage V_1 of: 10 + j17.32

The resultant voltage, V_T is found by adding together the horizontal and vertical components as follows:

$V_{Horizontal}$ = sum of real parts of V_1 and V_2 = 30 + 10 = 40 volts

$V_{Vertical}$ = sum of imaginary parts of V_1 and V_2 = 0 + 17.32 = 17.32 volts

Now that both the real and imaginary values have been found the magnitude of voltage, V_T is determined by simply using Pythagoras's Theorem for a 90° triangle as follows:

$$V_T = \sqrt{\begin{pmatrix}\text{Real or Horizontal}\\ \text{Compoent}\end{pmatrix}^2 + \begin{pmatrix}\textit{Imaginary or Vertical}\\ \textit{Component}\end{pmatrix}^2}$$

$$V_T = \sqrt{40^2 + 17.32^2}$$

$$\therefore V_T = 43.6 \text{ volts}$$

Then the resulting phasor diagram will be:

Resultant Value of V_T

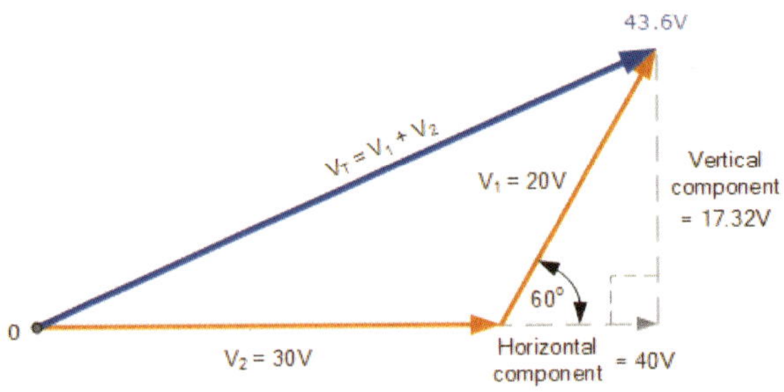

Phasor Subtraction

Phasor subtraction is very similar to the above rectangular method of addition, except this time the vector difference is the other diagonal of the parallelogram between the two voltages of V_1 and V_2 as shown.

Vector Subtraction of Two Phasors

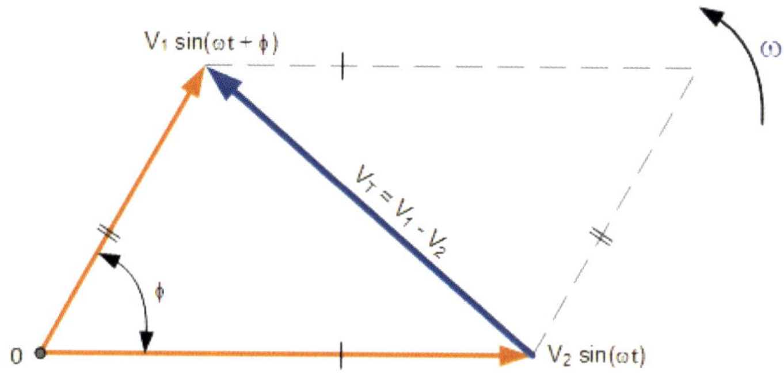

This time instead of "adding" together both the horizontal and vertical components we take them away, subtraction:

$$A=x+jy \qquad B=w+jz$$

$$A-B=(x-w)+j(y-z)$$

The 3-Phase Phasor Diagram

Previously we have only looked at single-phase AC waveforms where a single multi-turn coil rotates within a magnetic field. But if three identical coils each with the same number of coil turns are placed at an electrical angle of 120° to each other on the same rotor shaft, a three-phase voltage supply would be generated.

A balanced three-phase voltage supply consists of three individual sinusoidal voltages that are all equal in magnitude and frequency but are out-of-phase with each other by exactly 120° electrical degrees.

Standard practice is to colour code the three phases as Red, Yellow and Blue to identify each individual phase with the red phase as the reference phase. The normal sequence of rotation for a three phase supply is Red followed by Yellow followed by Blue, (R, Y, B).

As with the single-phase phasors above, the phasors representing a three-phase system also rotate in an anti-clockwise direction around a central point as indicated by the arrow marked ω in rad/s. The phasors for a three-phase balanced star or delta connected system are shown below.

Three-phase Phasor Diagram

The phase voltages are all equal in magnitude but only differ in their phase angle. The three windings of the coils are connected together at points, a_1, b_1 and c_1 to produce a common neutral connection for the three individual phases. Then if the red phase is taken as the reference phase each individual phase voltage can be defined with respect to the common neutral.

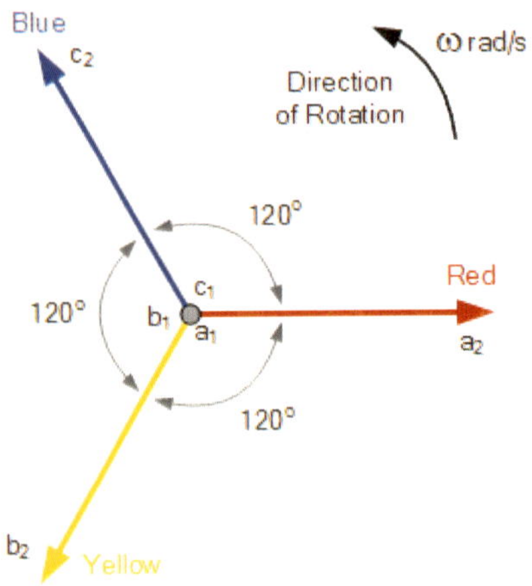

Three-phase Voltage Equations

Red Phase: $V_{RN} = V_m \sin\theta$

Yellow Phase: $V_{YN} = V_m \sin\left(\theta\text{-}120^\circ\right)$

Blue Phase: $V_{BN} = V_m \sin\left(\theta - 240^\circ\right)$

or

$$V_{BN} = V_m \sin\left(\theta + 120^\circ\right)$$

If the red phase voltage, V_{RN} is taken as the reference voltage as stated earlier then the phase sequence will be R – Y – B so the voltage in the yellow phase lags V_{RN} by 120°, and the voltage in the blue phase lags V_{YN} also by 120°. But we can also say the blue phase voltage, V_{BN} leads the red phase voltage, V_{RN} by 120°.

One final point about a three-phase system. As the three individual sinusoidal voltages have a fixed relationship between each other of 120° they are said to be "balanced" therefore, in a set of balanced three phase voltages their phasor sum will always be zero as: $V_a + V_b + V_c = 0$

References

- Electrical-dc-series-and-parallel-circuit: electrical4u.com, Retrieved 31 March 2018
- What-is-an-ac-circuit: circuitglobe.com, Retrieved 23 April 2018
- What-is-resistance, measurement-basics, best-practices: fluke.com, Retrieved 13 May 2018
- Electrical-resistance, Causes-of-resistance: newworldencyclopedia.org, Retrieved 23 March 2018
- Electrical-impedance-and-its-applications: elprocus.com, Retrieved 09 March 2018
- How-to-use-phasors-for-circuit-analysis, science-electronics: dummies.com, Retrieved 29 March 2018

5

Measuring and Testing Instruments

Some of the important instruments used in electronics are ammeters, ohmmeters, multimeters, oscilloscopes, logic probes, spectrum analyzers and signal generators. This chapter analyzes their working principles and their use in the measurement of physical quantities and testing in electronics.

Ammeter

Ammeter is the instrument for measuring either direct or alternating electric current, in amperes. An ammeter can measure a wide range of current values because at high values only a small portion of the current is directed through the meter mechanism; a shunt in parallel with the meter carries the major portion.

Ammeters vary in their operating principles and accuracies. The D'Arsonval-movement ammeter measures direct current with accuracies of from 0.1 to 2.0 percent. The electrodynamic ammeter uses a moving coil rotating in the field produced by a fixed coil. It measures direct and alternating currentwith accuracies of from 0.1 to 0.25 percent. In the thermal ammeter, used primarily to measure alternating current with accuracies of from 0.5 to 3 percent, the measured current heats a thermoconverter (thermocouple); the small voltage thus generated is used to power a millivoltmeter. Digital ammeters, with no moving parts, use a circuit such as the dual slope integrator to convert a measured analogue (continuous) current to its digital equivalent. Many digital ammeters have accuracies better than 0.1 percent.

The ideal ammeter has zero internal resistance. But practically the ammeter has small internal resistance. The measuring range of the ammeter depends on the value of resistance.

Symbolic Representation

The capital alphabet A represents the ammeter in the circuit.

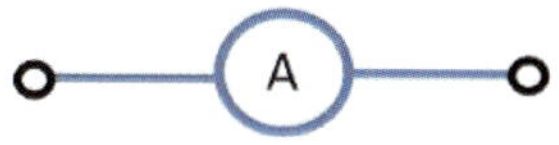

Connection of Ammeter in Circuit

The ammeter is connected in series with the circuit so that the whole electrons of measurand current passes through the ammeter. The power loss occurs in ammeter because of the measurand current and their internal resistance. The ammeter circuit has low resistance so that the small voltage drop occurs in the circuit.

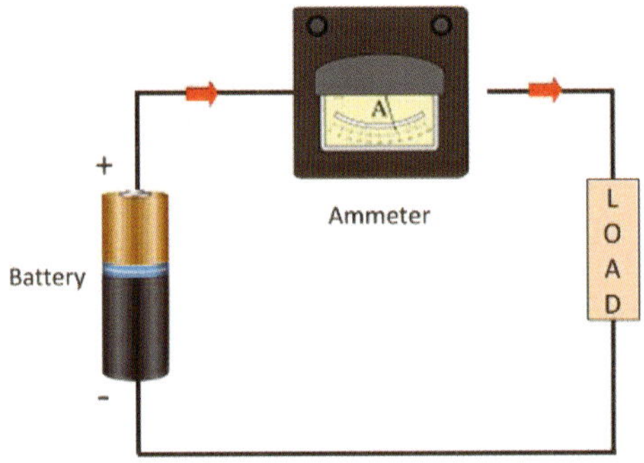

The resistance of the ammeter is kept low because of the two reasons:

- The whole measurand current passes through the ammeter.
- The low voltage drop occurs across the ammeter.

Also due to very low impedance the power loss will be low and if it is connected in parallel it becomes almost a short circuited path and all the current will flow through ammeter as a result of high current the instrument may burn. So due to this reason it must be connected in series. For an ideal ammeter, it must have zero impedance so that it has zero voltage drop across it so the power loss in the instrument is zero. But the ideal is not achievable practically.

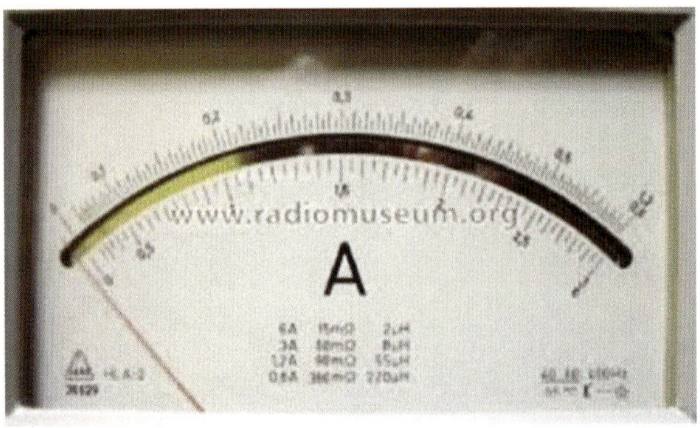

Classification or Types of Ammeter

Depending on the constructing principle, there are many types of ammeter we get, they are mainly:

1. Permanent Magnet Moving Coil(PMMC) ammeter.

2. Moving Iron(MI) Ammeter.
3. Electrodynamometer type Ammeter.
4. Rectifier type Ammeter.

Depending on this types of measurement:

1. DC Ammeter.
2. AC Ammeter.

DC Ammeter are mainly PMMC instruments, MI can measure both AC and DC currents, also Electrodynamometer type thermal instrument can measure DC and AC, induction meters are not generally used for ammeter construction due to their higher cost, inaccuracy in measurement.

Description of Different Types of Ammeters

PMMC Ammeter

Principle PMMC Ammeter: When current carrying conductor placed in a magnetic field, a mechanical force acts on the conductor, if it is attached to a moving system, with the coil movement, the pointer moves over the scale. Explanation: As the name suggests it has permanent magnets which are employed in this kind of measuring instruments. It is particularly suited for DC measurement because here deflection is proportional to the current and hence if current direction is reversed, deflection of the pointer will also be reversed so it is used only for DC measurement. This type of instrument is called D Arnsonval type instrument. It has major advantage of having linear scale, low power consumption, high accuracy. Major disadvantage of being measured only DC quantity, higher cost etc. Deflecting torque,

$$T \quad BiNlbNm$$

Where,

B = Flux density in Wb/m^2.

i = Current flowing through the coil in Amp.

l = Length of the coil in m.

b = Breadth of the coil in m.

N = No of turns in the coil.

Extension of Range in a PMMC Ammeter

Now it looks quite extraordinary that we can extend the range of measurement in this type of instrument. Many of us will think that we must buy a new ammeter to measure higher amount of current and also many of us may think we have to change the constructional feature so that we can measure higher currents, but there is nothing like that, we just have to connect a shunt resistance in parallel and the range of that instrument can be extended, this is a simple solution provided by the instrument.

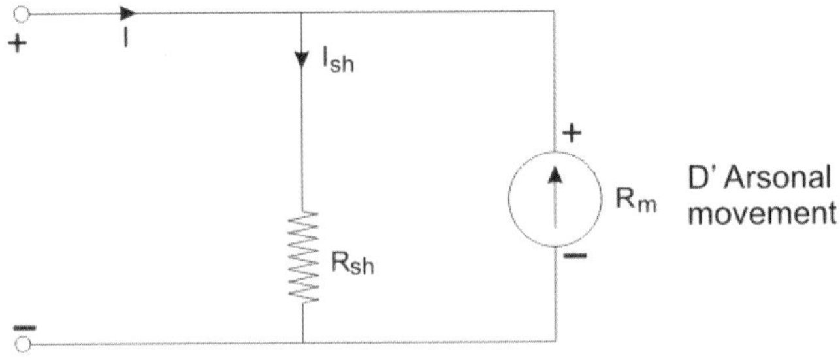

In the figure,

I = total current flowing in the circuit in Amp.

I_{sh} is the current through the shunt resistor in Amp.

R_m is the ammeter resistance in Ohm.

$$Then,\ R_{sh} = \frac{R_m}{\frac{I}{I - I_{sh}} - 1}$$

MI Ammeter

It is a moving iron instrument, used for both AC and DC, It can be used for both because the deflection θ proportional square of the current so whatever is the direction of current, it shows directional deflection, further they are classified in two more ways:

- Attraction type.
- Repulsion type.

Torque equation is:

$$T = \frac{1}{2} I^2 \frac{dL}{d\theta}$$

Where,

I is the total current flowing in the circuit in Amp.

L is the self inductance of the coil in Henry.

θ is the deflection in Radian.

- Attraction Type MI Instrument Principle: When an unmagnetised soft iron is placed in the magnetic field, it is attracted towards the coil, if a moving system attached and current is passed through a coil, it creates a magnetic field which attracts iron piece and creates deflecting torque as a result of which pointer moves over the scale.
- Repulsion Type MI Instrument Principle: When two iron pieces are magnetized with same polarity by passing a current than repulsion between them occurs and that repulsion produces a deflecting torque due to which the pointer moves. The advantages of MI instruments are they can measure both AC and DC, cheap, low friction errors, robustness etc. It is mainly used in AC measurement because in DC measurement error will be more due to hysteresis.

Electrodynamometer Type Ammeter

This can be used to measure both AC and DC currents. Now we see that we have PMMC and MI instrument for the measurement of AC and DC currents, a question may arise - "why do we need Electrodynamometer Ammeter? If we can measure current accurately by other instrument also?"

The answer is Electrodynamometer instruments have the same calibration for both AC and DC i.e. if it is calibrated with DC, then also without calibrating we can measure AC.

Principle Electrodynamometer Type Ammeter: There we have two coils, namely fixed and moving coils. If a current is passed through two coils it will stay in the zero position due to the development of equal and opposite torque. If somehow, the direction of one torque is reversed as the current in the coil reverses, an unidirectional torque is produced. For ammeter, the connection is a series one and $\varphi = 0$ Where, φ is the phase angle.

$$T = I^2 \frac{dM}{d\theta}$$

Where, I is the amount of current flowing in the circuit in Amp. M = Mutual inductance of the coil. They have no hysteresis error, used for both AC and DC measurement, the main disadvantages are they have low torque/weight ratio, high friction loss, expensive than other measuring instruments etc.

Rectifier Ammeter

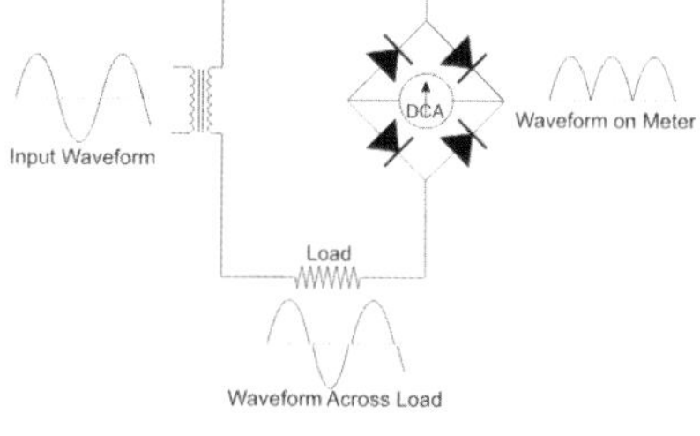

Principle of Rectifier Ammeter: They are used for AC measurement which is connected to secondary of a current transformer, the secondary current is much less than primary and connected with a bridge rectifier to a moving coil ammeter.

Advantages

1. It can be used in high frequency also.
2. Uniform scale for most of the ranges.

Disadvantages being error due to temperature decrease in sensitivity in AC operation.

Ammeter Shunt

The high-value current directly passes through the ammeter which damages their internal circuit. For removing this problem, the shunt resistance is connected in parallel with the ammeter.

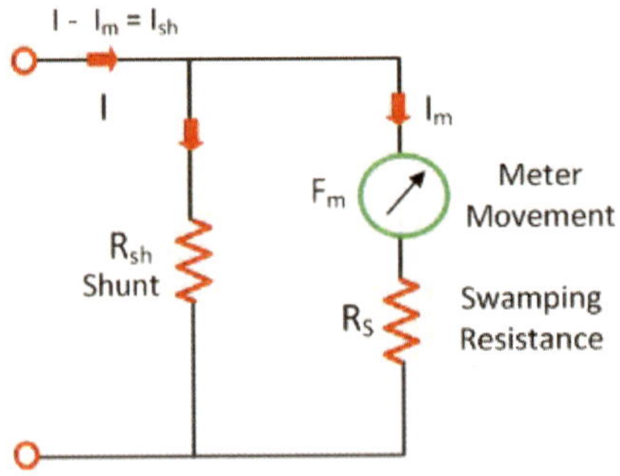

Meter shunt and swamping resistance

If the large measurand current passes through the circuit, the major portion of the current passes through the shunt resistance. The shunt resistance will not affect the working of the ammeter, i.e., the movement of the coil remains same.

Effect of Temperature in Ammeter

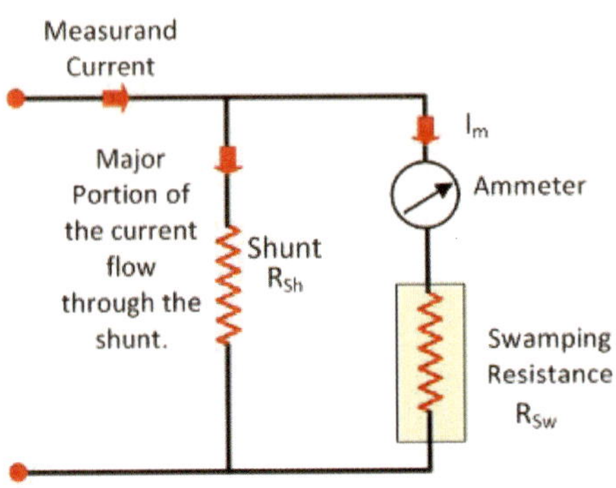

Swamping resistance

The ammeter is a sensitive device which is easily affected by the surrounding temperature. The variation in temperature causes the error in the reading. This can reduce by swamping resistance. The resistance having zero temperature coefficient is known as the swamping resistance. It connects in series with the ammeter. The swamping resistance reduces the effect of temperature on the meter.

The ammeter has the inbuilt fuse which protects the ammeter from the heavy current. If substantial current flows through the ammeter, the fuse will break. The ammeter is not able to measure the current until the new one does not replace the fuse.

Ohmmeters

The meter which measures the resistance and the continuityof the electrical circuit and their components such type of meter is known as the ohmmeter. It measures the resistance in ohms. The micro-ohmmeter is used for measuring the low resistance and the mega ohmmeter measures the high resistance of the circuit. The ohmmeter is very convenient to use but less accurate.

Ohmmeter Working

In an Ohmmeter, the deflection of the needle is controlled by the amount of battery current. Before calculating the resistance of an unknown electrical circuit or resistor, first of all, the test leads of the Ohmmeter are shorted together.

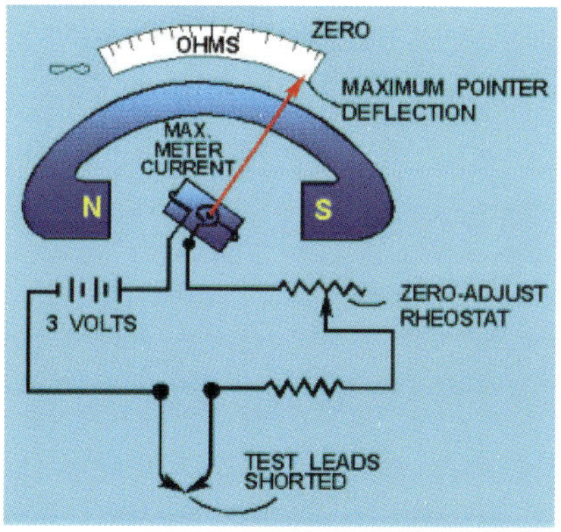

A simple ohmmeter circuit

When the leads are shorted, the meter is adjusted for proper operation on the selected range and the needle drives back to the maximum position on the Ohms scale and the meter current is max. After using an Ohmmeter, the test leads should be removed. If the test leads remain connected to the Ohmmeter, then the battery of the meter gets discharged. When the rheostat is adjusted properly, with the test leads shorted, the needle of the meter comes to zero position, and this specifies a zero resistance between the test leads.

Measuring Circuit Working with an Ohmmeter

When this meter is adjusted for zero reading on the Ohms scale, it is ready for measuring resistance in a circuit. The arrangement of Ohmmeter with a typical circuit is shown below. The power switch of the circuit should always be in the off position. Because the source voltage being applied to the circuit across the meter could damage the meter.

When the test leads of the Ohmmeter are connected serially to the circuit, this causes the current to flow through the circuit being tested. If the meter's test leads are connected at a point a & b in the circuit, then the amount of the current in the meter coil will depend on the resistance of the meter and the total resistance of the resistors R_1 and R_2. As the meter in the above circuit is pre adjusted one, the amount of coil movement therefore depends exclusively on the resistance of the R_1 & R_2. The addition of R_1 and R_2 raises the total series resistance, decreasing the current and therefore decreasing the needle deflection. The needle then comes to rest at a scale for indicating the combined resistance of the R_1 and R_2.

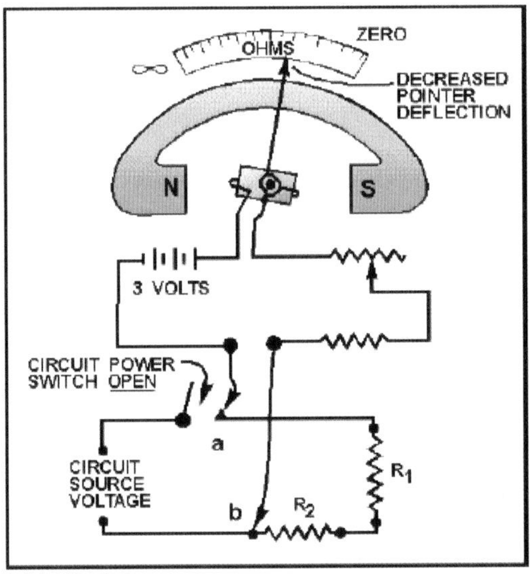

Circuit working with an ohmmeter

If the current limiting resistor R_1 or zero adjusts resistor R_2 or both are replaced with a larger value resistor, then the current flow and the deflection in the moving coil of the meter will be decreased.

D'Arsonval Movement

This type of movement is used in DC measuring instruments. The main principle in these types of instruments is that when a current carrying coil which is placed in a magnetic field, it feels a force and that force can deflect the pointer of a meter and we get the reading in the instrument.

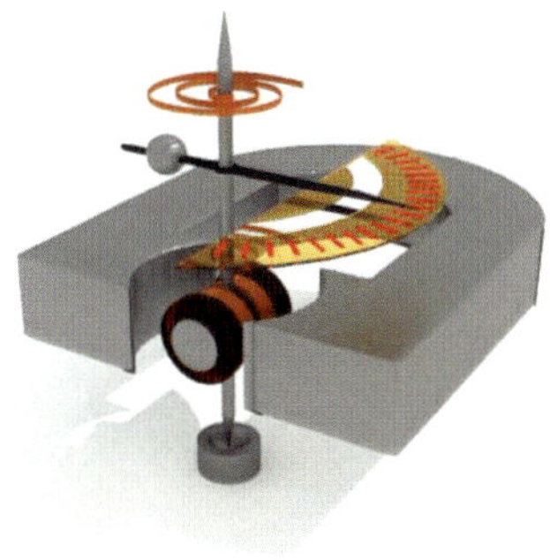

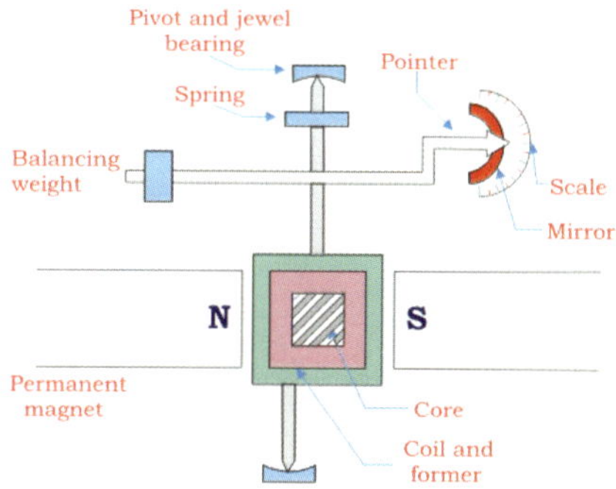

Construction of D'Arsonval Movement

This type of instruments consists of a permanent magnet and a coil which carries current and is placed in between them. The coil may be of rectangular or circular in shape. The iron core is used to provide flux of low reluctance so it produces high intensity magnetic field. Due to high intensity magnetic fields the deflecting torque produced is of large value due to which sensitivity of the meter is also increased. Current which entered comes out of two control springs, one in the upper side and one in the lower side. If the direction of current is reversed in these types of instruments, then torque direction will also be reversed so these types of instruments are applicable in DC measurements only. The deflecting torque is directly proportional to the deflection angle hence these types of instruments have the linear scale. To limit the deflection of the pointer we have to use damping which provides an equal and opposite force to the deflecting torque and hence the pointer comes to rest at a certain value. The indication of the breeding is given by a mirror in which a beam of light is reflected on to the scale and hence deflection can be measured. There are many advantages due to which we use D'Arsonval type instrument. They are:

- They have uniform scale.
- Effective eddy current damping.
- Low power consumption.
- No hysteresis loss.
- They are not affected by stray fields.

Owing to posses those major advantages we can use this type of instrument. However they suffer from drawbacks such as:

- It cannot be used in AC.
- Costlier compared to MI instruments.
- There may be error due to ageing of springs by which we may not get accurate result.

However in case of resistance measurement we go for DC measurement because of the advantages offered by PMMC instruments and we multiply that resistance by 1.6 to find out AC resistance, so these instruments are much widely used due to their advantages. The disadvantages offered by it are dominated by the advantages so they are used.

Types of Ohmmeter

The ohmmeter gives the approximate value of resistance. It is very portable and hence used in the laboratory. It is of three types; series ohmmeter, shunt ohmmeter and multi-range ohmmeter.

Series Ohmmeter

In series ohmmeter, the measuring resistance component or circuit is connected in series with the meter. The value of resistance is measured through the d'Arsonval movement connected in parallel with the shunt resistor R_2. The parallel resistance R_2 is connected in series with the resistance R_1 and the battery. The component whose resistance is used to be measured is connected in series with the terminal A and B.

The circuit diagram of the series ohmmeter is shown in the figure below:

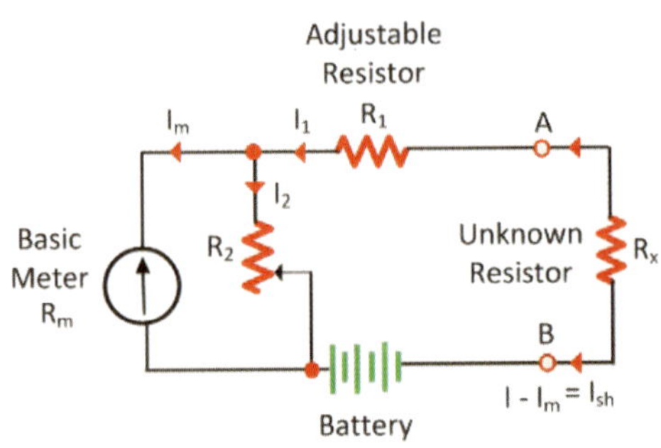

Series type ohmmeter

When the value of unknown resistance is zero the large current flow through the meter. In this condition, the shunt resistance is adjusted until the meter indicates the full load current. For full load current, the pointer deflects towards zero 0 ohms.

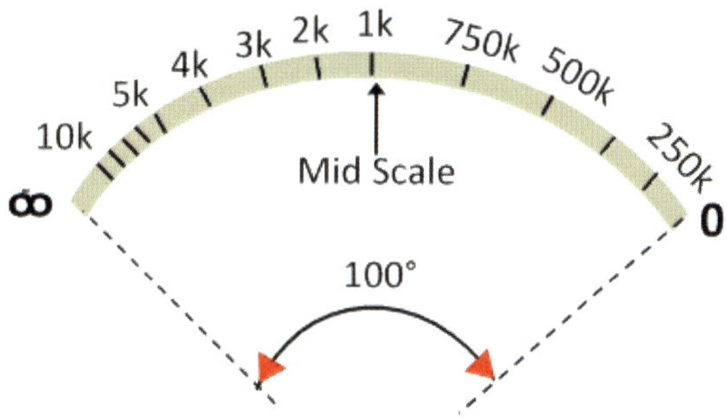

Scale of series type ohmmeter

When the unknown resistance R_x is removed from the circuit the resistance of the circuit becomes infinite and no current flow through the circuit. The pointer of the meter deflects towards the ∞ (infinity). The meter shows the infinite resistance at zero current and the zero resistance when full range current flows through it.

When the unknown resistance is connected in series with the circuit and if their resistance is high, then the pointer of the meter deflects toward the left. And if the resistance is low, then pointer deflects toward the right.

Shunt Type Ohmmeter

The meter in which the measuring resistance is connected in parallel with the battery is known as the shunt ohmmeter. It is mainly used for measuring the low-value resistance.

The circuit diagram of the shunt ohmmeter is shown in the figure below:

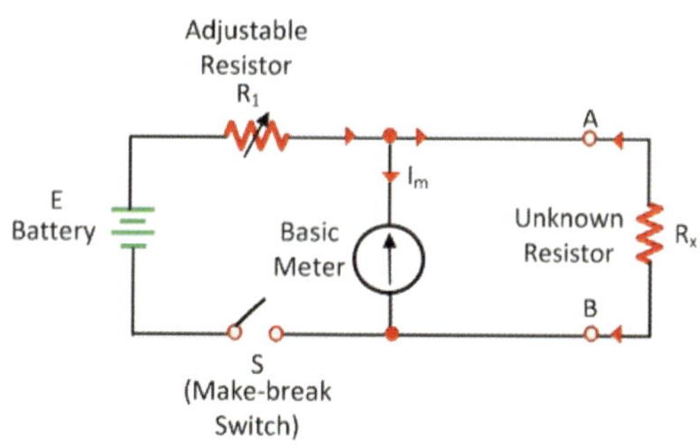

Shunt type ammeter

The battery (E), basic meter (R_m) and the adjustable resistance are the main components of the shunt ohmmeter. The unknown resistance is connected across terminal A and B.

When the value of unknown resistance is zero the meter current becomes zero. And if the resistance becomes infinite (i.e., the terminal A and B are open) then the current passes through the battery and the pointer shows the full-scale deflection toward left.

The shunt type ohmmeter has the zero mark (no current) on the left of the scale and the infinity mark on their right side.

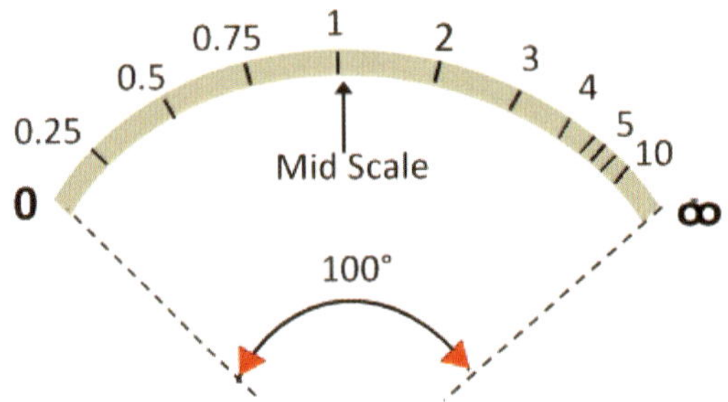

Scale of shunt type ohmmeter

Multi-Range Ohmmeter

The range of this type of ohmmeter is very high. The meter has adjuster which selects the range according to need.

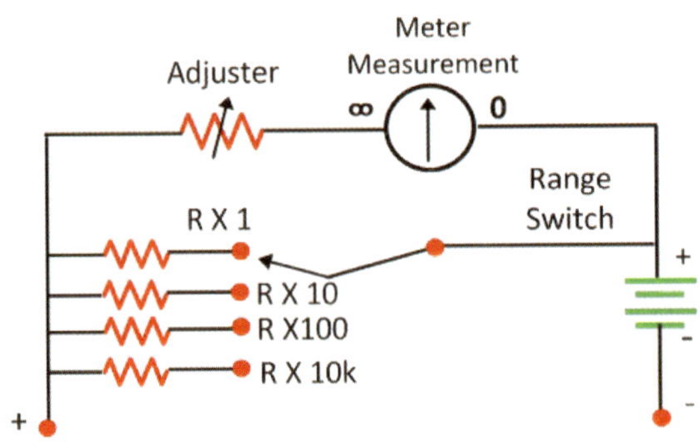

Multi-Range Ohmmeter

For example, consider we use the meter for measuring the resistance under 10 ohms. For this first, we have to set the range of 10 ohms. The resistance whose value is used to be measured is connected in parallel with the meter. The magnitude of the resistance is determined through the deflection of the pointer.

Applications of Ohmmeter

- These are being widely used now a days to check circuit continuity, i.e. if enough current is flowing through the circuit or there is infinite resistance between two points and the circuit id disconnected.
- They are also used as laboratory testing equipment in different experiments and for teaching purposes.
- They are quite useful in debugging small chips like printed circuit boards and other stuff which needs to be implemented in sensitive equipment.

Multimeters

A multimeter is a handy tool that you use to measure electricity, just like you would use a ruler to measure distance, a stopwatch to measure time, or a scale to measure weight. The neat thing about a multimeter is that unlike a ruler, watch, or scale, it can measure different things — kind of like a multi-tool. Most multimeters have a knob on the front that lets you select what you want to measure. Below is a figure of a typical multimeter:

Figure: A typical multimeter.

Measurments done using Multimeters

Almost all multimeters can measure voltage, current, and resistance.

Some multimeters have a continuity check, resulting in a loud beep if two things are electrically connected. This is helpful if, for instance, you are building a circuit and connecting wires or soldering; the beep indicates everything is connected and nothing has come loose. You can also use it to make sure two things are *not* connected, to help prevent short circuits.

Some multimeters also have a diode check function. A diode is like a one-way valve that only lets electricity flow in one direction. The exact function of the diode check can vary from multimeter to multimeter. If you're working with a diode and can't tell which way it goes in the circuit, or if you're not sure the diode is working properly, the check feature can be quite handy. If your multimeter has a diode check function, read the manual to find out exactly how it works.

Advanced multimeters might have other functions, such as the ability to measure and identify other electrical components, like transistors or capacitors.

Most multimeters also use metric prefixes. Metric prefixes work the same way with units of electricity as they do with other units you might be more familiar with, like distance and mass. For example, you probably know that a meter is a unit of distance, a kilometer is one thousand meters, and a millimeter is one thousandth of a meter. The same applies to milligrams, grams, and kilograms for mass. Here are the common metric prefixes you will find on most multimeters.

- μ (micro): one millionth
- m (milli): one thousandth
- k (kilo): one thousand
- M: (mega): one million

These metric prefixes are used in the same way for volts, amps, and ohms. For example, 200kΩ is pronounced "two hundred kilo-ohms," and means two hundred thousand (200,000) ohms.

Some multimeters are "auto-ranging," whereas others require you to manually select the range for your measurement. If you need to manually select the range, you should always pick a value that is *slightly higher* than the value you expect to measure. Think about it like using a ruler and a yardstick. If you need to measure something that is 18 inches long, a 12-inch ruler will be too short; you need to use the yardstick. The same applies to using a multimeter. Say you are going to measure the voltage of a AA battery, which you expect to be 1.5V. The multimeter on the left in figure below has options for 200mV, 2V, 20V, 200V, and 600V (for direct current). 200mV is too small, so you would pick the next highest value that works: 2V. All of the other options are unnecessarily large, and would result in a loss in accuracy (it would be like using a 50-foot tape measure that only has markings every foot, and no inch markings; it isn't as accurate as using a yardstick with 1-inch markings).

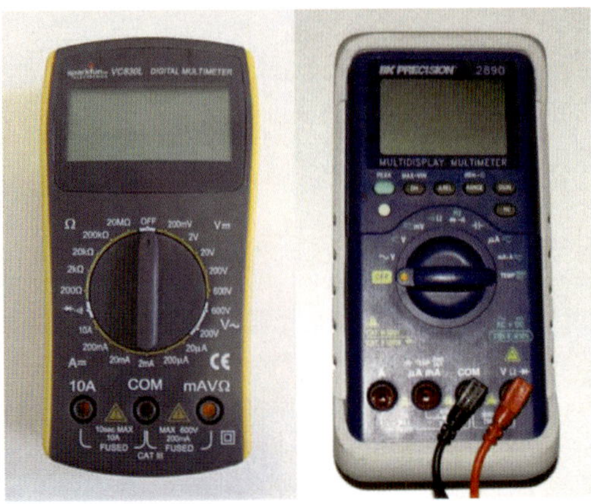

Figure: The multimeter on the left is manual-ranging, with many different options (indicated by metric prefixes) for measuring different amounts of voltage, current, and resistance. The multimeter on the right is auto-ranging, meaning it will automatically select the appropriate range.

DMM Facilities

While the facilities that a digital multimeter can offer are much greater than their analogue predecessors, the cost of DMMs is relatively low. DMMs are able to offer as standard the basic measurements that would typically include:

- Current (DC)
- Current (AC)
- Voltage (DC)
- Voltage (AC)
- Resistance

However, using integrated circuit technology, most DMMs are able to offer additional test capabilities. These may include some of the following:

- Capacitance
- Temperature
- Frequency
- Transistor test - hfe, etc
- Continuity (buzzer)

While some of these additional test features may not be as accurate as those supplied by dedicated test instruments, they are nevertheless very useful, especially where approximate readings only are needed.

In addition to an increase in the number of basic measurements that can be made, refinements of some of the basic measurements are also available on some models. True RMS multimeters are available. In many instances, AC waveforms use forms of average measurements that are then converted to RMS measurements using a form factor. This method of measurement is very dependent upon the shape of the waveform and as a result a true RMS digital multimeter may be required. In addition to the availability of a true RMS meters, similar refinements of the other basic measurements are also available in some instances.

Multimeter Selection and Functions

To choose a DMM that is best for your application, first consider resolution, digit and accuracy.

Resolution refers to how fine a measurement a meter can make. By knowing the resolution of a meter you can determine if it is possible to see a small change in the measured signal. The terms digits and counts are used to describe a meter's resolution. DMMs are grouped by the number of counts or digits they display. A 3½-digit meter can display three full digits ranging from 0-9 and one half digit which displays only a "1" or is left blank. A 3½ digit meter will display up to 1,999 counts of resolution and a 4½ digit meter can display up to 19,999 counts of resolution. Some DMMs may have enhanced resolution offering higher counts within their digit range.

Accuracy is the largest allowable error in the readings. It's an indication of how close the DMM's displayed measurement is to the actual value of the signal being measured. Accuracy is generally expressed as a percent of the reading. An accuracy of one percent of the reading means that for a display of 100 volts, the actual value of the voltage could be somewhere between 99 and 101 volts.

Autoranging, or dialing, is also a useful multimeter feature. It allows the user to quickly move from the instruments various measuring scales while maintaining accurate readings.

Another important consideration to take into account when selecting a DMM is whether it is a True RMS (root mean square) meter. The typical DMM is not a True RMS meter and as a result it will produce misleading voltage readings when used to measure anything other than DC signals or pure sine wave AC signals. True RMS meters allow non-sinusoidal AC signals to be accurately measured.

Fundamentals of the working of a DMM

The key process that occurs within a digital multimeter for any measurement that takes place is that of voltage measurement. All other measurements are derived from this basic measurement.

Accordingly the key to understanding how a digital multimeter works is in understanding this process.

There are many forms of analogue to digital converter, ADC. However the one that is most widely used in digital multimeters, DMMs is known as the successive approximation register or SAR. Some SAR ADCs may only have resolution levels of 12 bits, but those used in test equipment including DMMs generally have 16 bits or possibly more dependent upon the application. Typically for DMMs resolution levels of 16 bits are generally used, with speeds of 100k samples per second. These levels of speed are more than adequate for most DMM applications, where high levels of speed are not normally required.

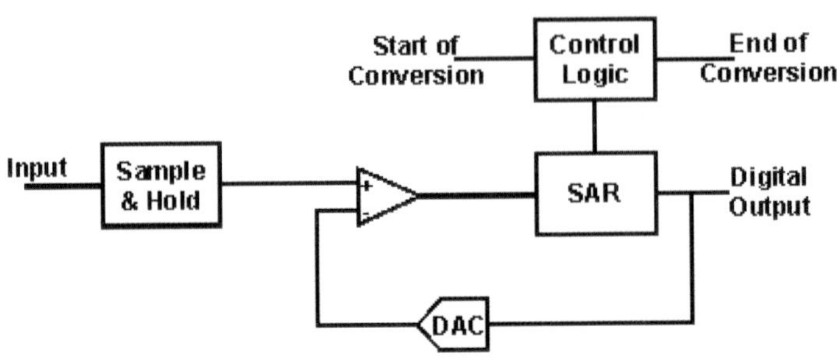

Successive approximation register ADC used in most DMMs

As the name implies, the successive approximation register ADC operates by successively homing in on the value of the incoming voltage.

The first stage of the process is for the sample and hold circuit to sample the voltage at the input of the DMM and then to hold it steady.

With a steady input voltage the register starts at half its full scale value. This would typically require the most significant bit, MSB set to "1" and all the remaining ones set to "0". Assuming that the input voltage could be anywhere in the range, the mid-range means that the ADC is set in the middle of the range and this provides a faster settling time. As it only has to move a maximum of the full scale rather than possibly 100%.

To see how it works take the simple example of an 4-bit SAR. Its output will start at 1000. If the voltage is less than half the maximum capability the comparator output will be low and that will force the register to a level of 0100. If the voltage is above this, the register will move to 0110, and so forth until it homes in on the nearest value.

It can be seen that SAR converters, need one approximating cycle for each output bit, i.e. an n-bit ADC will require n cycles.

DMM Operation

Although the analogue to digital converter forms the key element within the instrument, in order to fully understand how a digital multimeter works, it is necessary to look at some of the other functions around the ADC.

Although the ADC will take very many samples the overall digital multimeter will not display or return every sample taken. Instead the samples are buffered and averaged to achieve high accuracy and resolution. This will overcome the effects of small variations such as noise, etc., noise created by the analogue first stages of the DMM being an important factor that needs to be overcome to achieve the highest accuracy.

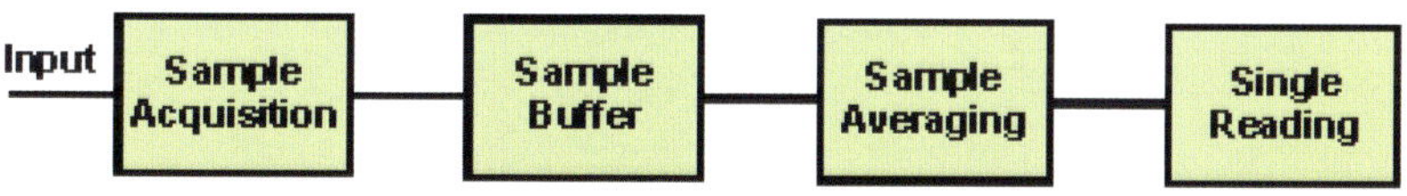

Operation flow diagram for operation of a DMM

Measurement Time

One of the key areas of understanding how a digital multimeter works is related to the measurement time. Apart from the basic measurement there are a number of other functions that are required and these all take time. Accordingly the measurement time of a digital multimeter, DMM, may not always appear straightforward.

The overall measurement time for a DMM is made up from several phases where different activities occur:

- Switch time: The switch time is the time required for the instrument to settle after the input has been switched. This includes the time to settle after a measurement type has been changed, e.g. from voltage to resistance, etc. It also includes time to settle after the range has been changed. If auto-ranging is included the meter will need to settle if a range change is required.
- Settling time: Once the value to be measured has been applied to the input, a certain time will be required for it to settle. This will overcome any input capacitance levels when high impedance tests are made, or generally for the circuit and instrument to settle.
- Signal measurement time: This is the basic time required to make the measurement itself. For AC measurements, the frequency of operation must be taken into account because the minimum signal measurement time is based on the minimum frequency required of the measurement. For example, for a minimum frequency of 50 Hz, an aperture of four time the period is required, i.e. 80 ms for a 50Hz signal, or 67ms for a 60Hz signal, etc.
- Auto-zero time: When autorange is selected, or range changes are made, it is necessary to zero the meter to ensure accuracy. Once the correct range is selected, the auto-zero is performance for that range.
- ADC calibration time: In some DMMs a calibration is periodically performed. This must be accounted for, especially where measurements are taken under automatic or computer control.

Oscilloscope

An oscilloscope (or CRO) is an instrument for observing electrical signals and is essential for anyone designing or repairing electronic equipment. The primary advantage of using an oscilloscope instead of a digital voltmeter is that the signal can be viewed directly on a screen, facilitating direct measurement of AC characteristics such as frequency, peak to peak (p-p) voltage and phase angle.

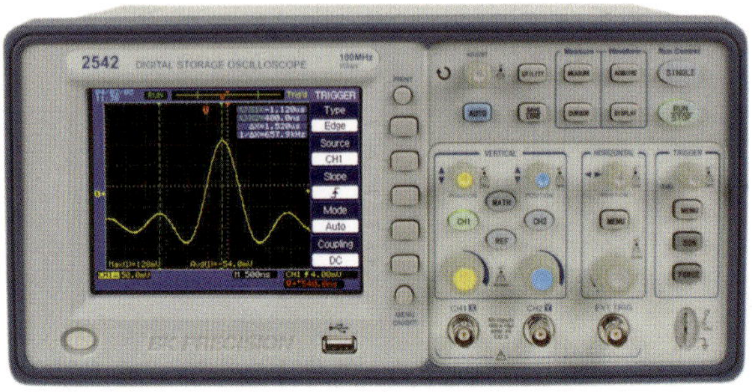

Figure: Oscilloscope used in the laboratory

A photograph of the oscilloscope used in this laboratory is shown in figure. It is capable of displaying two distinct external signals in two channels, labelled CH1 and CH2. We use a waveform generator as the ac source, shown in figure above. Turn on both units and allow them to warm up for a few minutes before proceeding. A photograph of the oscilloscopes used in this laboratory is shown in shown in figure above. They are capable of displaying two distinct external signals in two channels, labelled CH1 and CH2. We use a waveform generator as an ac voltage source, shown in shown in figure below. Both items are used extensively in ac circuits. Turn on both units and allow them to warm up for a few minutes before proceeding.

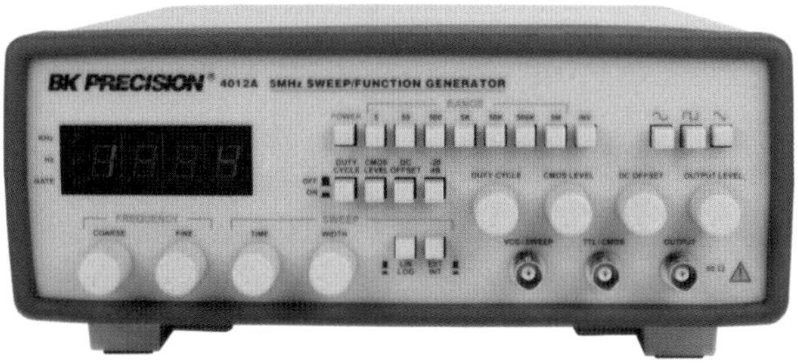

Figure: Waveform generator used in the laboratory.

Oscilloscope Measurements

The basic concept of the oscilloscope is that it displays waveforms in a two dimensional format. The vertical axis is normally used to plot the incoming voltage, and the horizontal axis is normally used as a time axis. In this way the waveform voltage can be displayed as a function of time. In this way a sine wave would be displayed in its well known graphical format with a line undulating line with time on the horizontal axis. This is the most common way of using an oscilloscope.

While it is useful to display a single waveform, many measurements and measurement techniques require more than on waveform to be displayed at any one time. This can be used to compare waveforms and look at how different waveforms interact or when events occur on more than one waveform. Many dual channel or multi-channel oscilloscopes exist that allow two or more waveforms to be displayed at the same time.

While the most usual way of displaying waveforms is to measure the instantaneous voltage on the vertical axis, and time on the horizontal axis, it is also possible to plot tow voltages against each other. In this way it is possible to display a Lissajous figure and look at the relative phases and voltages of two waveforms together.

At certain times it may be advantageous to be able to display three dimensional images. The traditional way of achieving this was to modulate the intensity of the image by using a third waveform. There is often a “Z Axis” input at the rear of many oscilloscopes to achieve this.

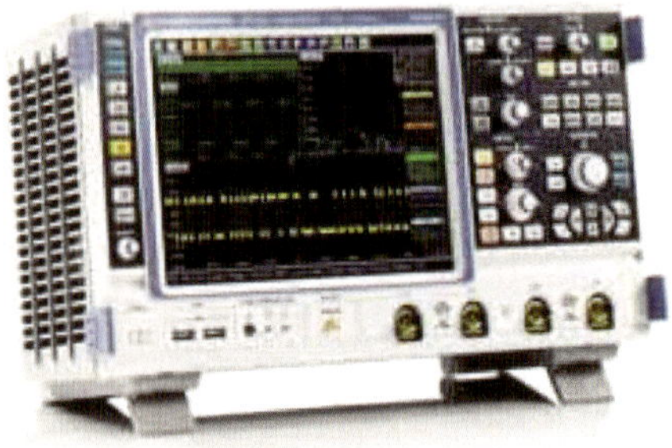

Display a Signal on the Oscilloscope

1. Connect a BNC cable from the OUTPUT terminal of the AC signal generator to channel CH1. Set the signal generator to output a sine wave with a frequency of a few hundred Hz.
2. Press the AUTO key, The oscilloscope automatically sets the vertical, horizontal, and trigger controls so that about three complete cycles of the waveform are displayed. You can adjust any of these manually using the Vertical and Horizontal controls if you need to optimize the display:
    - Position: Moves the trace for a particular channel vertically or horizontally on the screen.
    - Vertical Sensitivity: this control sets the voltage required to deflect the beam through one vertical division. The range for this CRO is from 2 mV/div up to 5 V/div. The sensitivity is displayed at the bottom left corner of the screen.
    - Time Base: The time base setting sets the time required for the beam to sweep across one large horizontal division. The vertical sensitivity and time are displayed at the bottom of the screen.

Types of Oscilloscope

Since the first oscilloscopes were developed and produced a variety of different types of oscilloscope have been developed. These have appeared as the technology has developed and they have been focused on particular applications and measurements. As a result, when choosing an oscilloscope, one of the first decisions to be made is regarding the basic form of oscilloscope that is needed.

Some of the major forms of oscilloscope are summarized in the list below.

- Analog oscilloscope
- Analog storage oscilloscope
- Digital oscilloscope

- Digital storage oscilloscope
- Digital phosphor oscilloscope
- Digital sampling oscilloscope

Oscilloscopes have been in use within the electronics industry for many years. Even with many new developments occurring and a shift to greater levels of software within products there is no lessening of the importance of oscilloscopes. As technology develops, so the technology used within scopes has enabled them to provide higher levels of performance and to provide new and useful functions. With the continuing movement of technology, oscilloscope technology will also move forwards.

Spectrum Analyzer

A spectrum analyzer is a device that displays signal amplitude (strength) as it varies by signal frequency. The frequency appears on the horizontal axis, and the amplitude is displayed on the vertical axis. To the casual observer, a spectrum analyzer looks like an oscilloscope, and in fact, some devices can function either as oscilloscopes or spectrum analyzers.

The electronics industry uses spectrum analyzers to examine the frequency spectrum of radio frequency (RF) and audio signals. These devices display the individual elements of these signals, as well as the performance of the circuits producing them. Through the use of spectrum analyzers, organizations can determine what modifications may be needed to reduce interference and thus improve the performance of Wi-Fi systems and wireless routers.

Working of Spectrum Analyzers

Most spectrum analyzers offer users the opportunity to set a start, stop and center frequency. The center frequency is halfway between the stop and start frequencies and is also the axis for the frequency used to determine the span -- the range between the start and stop frequencies. With an RF spectrum analyzer, the analyzer measures the radio "noise floor" and measures how close two signals can be while still being resolved into two separate peaks.

Features

Today, as spectrum analysis software and digital or spectrum analyzer app offerings have become more common, more analyzers are able to do analog-to-digital conversion and sample a significant input signal and frequency range. A modern spectrum analyzer may be able to show displayed average noise level, calculating the average noise level detected by the device. These detectors are typically capable of sample detection, peak detection or average detection.

Radio frequency spectrum bands

DESIGNATION	ABBREVIATION	FREQUENCIES	FREE-SPACE WAVELENGTHS
Very low frequency	VLF	9 kHz to 30 kHz	33 km to 10 km
Low frequency	LF	30 kHz to 300 kHz	10 km to 1 km
Medium frequency	MF	300 kHz to 3 MHz	1 km to 100 m
High frequency	HF	3 MHz to 30 MHz	100 m to 10 m
Very high frequency	VHF	30 MHz to 300 MHz	10 m to 1 m
Ultrahigh frequency	UHF	300 MHz to 3 GHz	1 m to 100 mm
Super-high frequency	SHF	3 GHz to 30 GHz	100 mm to 10 mm
Extremely high frequency	EHF	30 GHz to 300 GHz	10 mm to 1 mm

Uses for Spectrum Analyzers

A spectrum analyzer can be used to determine whether or not a wireless transmitter is working according to federally defined standards for purity of emissions. Output signals at frequencies other than the intended communications frequency appear as vertical lines (pips) on the display. A spectrum analyzer can also be used to determine, by direct observation, the bandwidth of a digital or analog signal.

Types of Spectrum Analyzers

Swept or Superheterodyne

A swept-tuned, or superheterodyne, spectrum analyzer down-converts part of the input signal to the center frequency of a band-pass filter by running a voltage-controlled oscillator across a range of frequencies. This allows for the full frequency range of the device to be analyzed. In this case, the resolution bandwidth is closely related to the minimum bandwidth detectable by the device and is controlled by the band-pass filter. The smaller the bandwidth, the greater the spectral resolution.

Fast Fourier Transform, FFT

Some digital spectrum analyzers use Fourier transforms, a way of decomposing a signal into its individual frequencies. These analyzers need a sampling frequency at least twice the bandwidth because frequency resolution is the inverse of the time over which the wave is measured and Fourier transformed.

Real-time

Real-time analyzers collect real-time bandwidth and sample the incoming RF spectrum in a limited span of time, converting the information using the fast Fourier transform (FFT) algorithm. Because it's real-time data collection, there is no "blind time," and there are no gaps in the calculated RF spectrum.

Audio

Spectrum analyzers can also be used in the audio spectrum, displaying volume levels of frequency bands audible to humans. This method is aimed at analyzing the harmonics of an audio signal. Once known as wave analyzers, these types of spectrum analyzers are widely used by sound engineers and can run on almost any computer equipped with a sound card.

Advantages and Disadvantages

Swept-tuned spectrum analyzers face tradeoffs between how rapidly the display can update a full frequency span and the resolution. Sometimes, if engineers are working with a very weak signal, a preamplifier is needed before analysis can begin. On the other hand, FFT analyzers can strain the capabilities of analog-to-digital converters and require significant processing power, limiting the possible frequency range that can be analyzed. Real-time FFT analyzers can offer good resolution while reducing potential gaps in sampling.

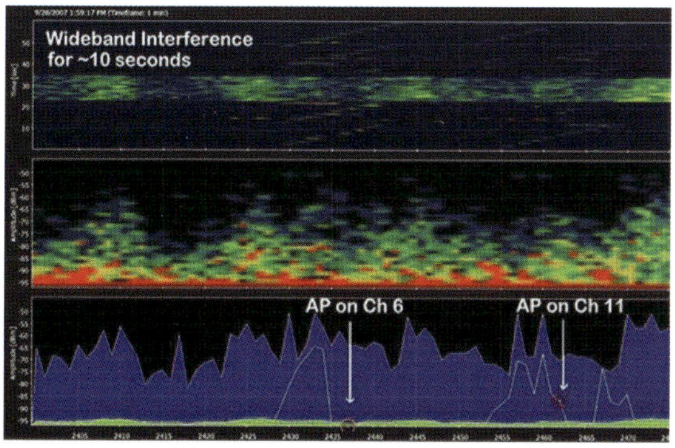

Spectrum analyzer visualizing airwaves

Spectrum Analyzer Interface

A spectrum analyzer interface is a device that can be connected to a wireless receiver or a personal computer to allow visual detection and analysis of electromagnetic signals over a defined band of frequencies. This is called panoramic reception, and it can be used to determine the frequencies of sources of interference to wireless networking equipment, such as Wi-Fi and wireless routers.

Signal Generator

The signal generator is an electronic device which induces the repeating and non-repeating analogue or digital signals. The signal generator is extensively used in troubleshooting, signal tracing, testing, adjustment of amplifier response and alignment of radio and T.V circuit. The signal generator has inbuilt modulation property. The amplitude and modulation property are very common in signal generators.

The block diagram of the signal generator is shown in the figure below. The main part of the signal generator is a VCO (voltage controlled oscillator). The input control voltage determined the frequency of the VCO.

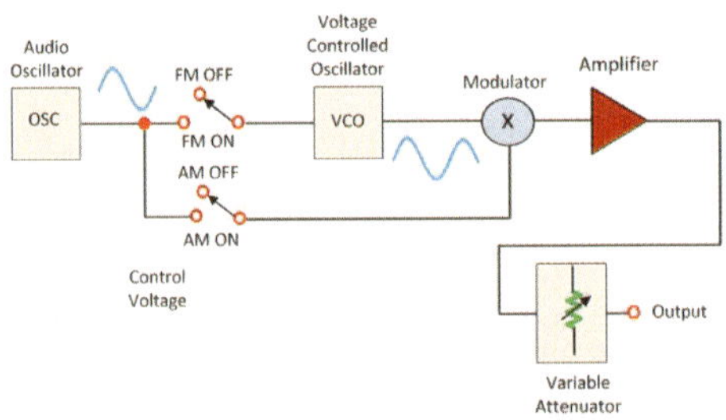

Block diagram of the signal generator

The frequency of the VCO is directly proportional to the control voltage. The signal applied to the control input gives the frequency of the oscillator. The frequency modulated signal is produced by the VCO when the audio input signal is applied across the control voltage.

The signal generator is also known as the generator which produces the tone, arbitrary and the digital pattern waveforms. The signal generators generate the modifies output signal along with the other signal which is the main difference between the signal generator and the oscillator.

When the signal produces the unmodulated signals, then they are said to be generating continuous height wave signal. It produces a modulated signal in the form of square waves, triangular waves, complex signals, etc.

For frequency modulation, the modulator circuit is placed after the VCO. The circuit changes the VCO output voltages by producing the output AM signal.

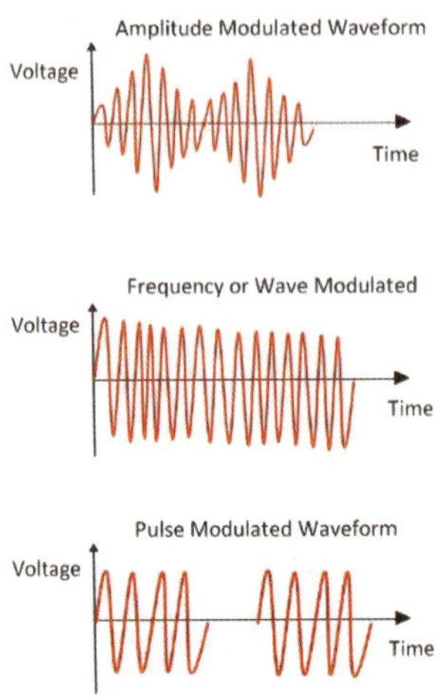

The accuracy and stability are the two main property of the signals which is used to test the receiver.

Types of Signal Generators

Functional Signal Generators

A function generator contains an electronic device called electronic oscillator which generates simple repetitive wave forms such as sine waves, square waves, triangular wave forms, and sawtooth wave forms. In modern devices, these wave forms are created by digital signal processing technique followed by analogous signals of lower frequencies so they are often required. A variety of function generators are black boxes with USB interfaces, used in instrumentation bus, and some are in the form of software. Their common uses are in the field of education, repairing electrical and electronic equipment and stimulus testing.

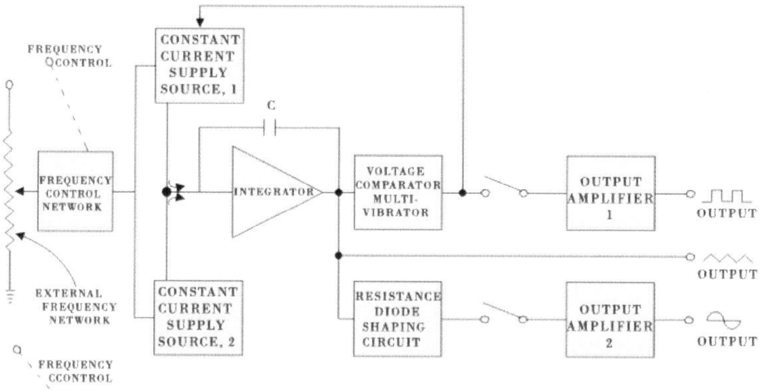

Block diagram of function generator

Arbitrary Waveform Generators

Arbitrary generators are a device that generates arbitrary streams of digital information. These waveforms do not have any fixed shape, and can be entered in a variety of waveforms. This is a generator with two independent output channels, which can stimulate two systems simultaneously. A common use of an arbitrary waveform generator is to stimulate a system with a complex waveform. It has a display screen that shows an exact picture of the arbitrary waveforms that helps in avoiding the chance of mistake while selecting waveform from memory. These types of generators have limited bandwidth and costlier than function generators. It is used in the semiconductor component, communication, and system tests.

RF Signal Generators

The radio frequency signal generator uses the variety of method to produce the signal like phase locked loop and direct digital synthesis etc. But most of the generators use frequency locked loop techniques to provide the stability and accuracy required by the system. It produces continuous wave tones of output frequency within their frequency range. The features of RF and microwave generators are almost same except they have different frequency range. Microwave signal generators have much wider frequency range as compared to RF signal generators. These type of generators are used for the testing system, audio and video broadcasting, satellite communications, radar and electronic warfare. Radio frequency generators are categorized in basically in three forms:

Analog Signal Generators

It is based on sine wave oscillator with a sharp distinction in design of radio frequency and audio frequency signal generators. But now it is outdated and digital electronics are in use.

Vector Signal Generators

Vector signal generators also called digital signal generators are capable of generating digitally modulated radio signals with complex modulation formats such as QPSK< QAM, etc.

Logical Signal Generators

Logical signal generators produce logic pulses in the form of conventional voltage levels. These generators are often called digital pattern generators. Pulse generators are able to generate pulses with variable delays and some even offer variable rise and fall times. It is used for functional validation and testing.

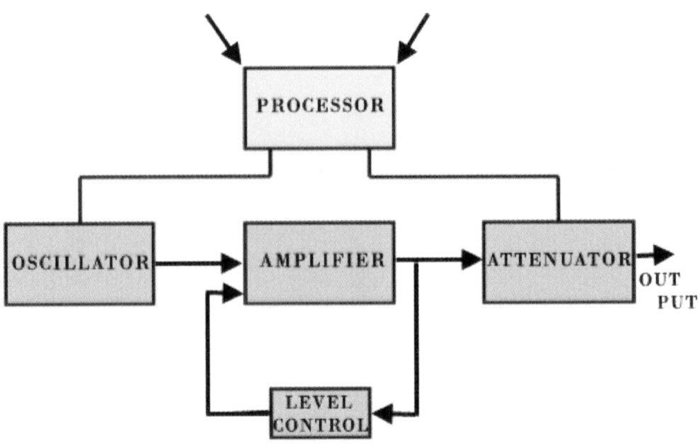

Audio Signal Generators

Audio signal generators generate signals in the range of audible i.e. typically in the range of 20 Hz to 20 KHz. It is used in checking frequency response in an audio system and the distortion measurements. Even a very low distortion could be measured with a relatively simple circuit by an audio signal generator. They have very low levels of harmonic distortion. These generators have a wide variety of application in electronic laboratories. Those generators which use intricate methods to generate audio signals for music are termed as the synthesizer.

Video Signal Generators

A video signal generator is a device which produces video waveforms. Along with that, some other signals are also used to stimulate faults. The important factor which effects video image on television is synchronization. That is why the output waveform of video generators normally contains synchronization signals, which includes vertical and horizontal sync. It has the variety of applications and digital formats.

Logic Probe

The logic probe or digital tester is normally a low cost handheld probe contained within a pen-like tube with indicator lights to show the state of the line being probed.

Simple logic probe tester

Typically logic probes are used to test digital circuits like those using TTL or CMOS logic. They often have three indicator lights on the body to indicate the state of the line. As such logic probes are very basic forms of digital testers, only able to test the state of a single line, but they can be useful in many applications.

The logic probe is normally powered by the circuit under test, there are normally leads with crocodile/alligator clips that can be attached to the ground and supply of the circuit under test.

Logic Probe Measurements

A logic probe is restricted in the number of measurements it can make when compared to other test instruments, but it can nevertheless be used for a variety of digital measurements:

- Logic high state: The logic probe/digital logic tester is able to detect lines that are at the digital or logic high state. The logic probe will indicate this typically with an LED which is often coloured red.
- Logic low: The logic probe also is able to indicate a logic or digital low. A common

indication is with the use of a green coloured LED.

- Digital pulses: The logic probe may incorporate some form of pulse detection circuitry. When the line is active and being pulsed a third colour, possibly amber will be indicated. The logic probe may well incorporate circuitry to detect very short pulses and in this way indicate when the line is active. Sometimes the length of the pulses may be indicated by the brightness of the LED.
- Line tri-stated: Some logic probes may also be able to detect when a line has been put into a tristate option. This is when the output device has its output turned off and no real logical state is defined. Many logic probes are able to indicate this state and they may do this by having all indicators turned off.

Logic probes vary from one manufacturer the next and therefore it is necessary to check exactly what measurements can be made and how the results are indicated.

Logic Probe Connections

Before using the logic probe, it is necessary to understand the connections.

As can be seen, there are three connections to the logic probe:

1. Black lead with crocodile clip: There are two leads which generally come out of the opposite end of the instrument to the metal probe itself. The black lead is connected to negative ground and is also used as the return.
2. Red lead with crocodile clip: This lead will have red on it somewhere, possibly just on the crocodile/alligator clip and is used to connect to the supply. Be careful to connect this to the logic supply which will normally be +5 volts and for some CMOS families up to 15 volts. Read the instructions to see over what range the probe will operate using a voltage higher than that which is specified could damage the logic probe.
3. Probe: The probe, as shown on the diagram is a metal point used for probing the circuit. When using this, be careful that the probe cannot slip and cause a short circuit which could damage the circuit under test.

The first requirement before using the logic probe is to connect the power connections to the circuit. Apart from ensuring they are the right voltage, the points used for connecting the crocodile clips to should be accessible and provide a reliable connection without the risk of touching any adjacent components or other connections.

Advantages and Disadvantages of a Logic Probe

As with any item of test equipment, there are advantages and disadvantages to the use of a logic probe tester than need to be considered before buying or using one.

Logic probe advantages -	Logic probe disadvantages -
• Low cost: A logic probe does not contain much circuitry, and the display is very rudimentary. Therefore the cost of manufacture is very low – they can typically be bought for less than the cost of a very basic multimeter.	• Very rough measurement: The nature of the logic probe means that only an indication of the presence of a logic signal can be detected. It is not replacement for a test instrument such as an oscilloscope.
• Ease of use: To use a logic probe typically requires the connection of power leads and then connecting the probe to the required point on the circuit.	• Poor display: A logic probe only uses a few LEDs to indicate the nature of the logic signal. As a result, little information can be displayed about the nature of the logic signal that is detected.

A logic probe tester is a very cheap and simple item of test equipment. It is able to provide a quick but very basic test for many logic circuits. However it is not nearly as flexible as an oscilloscope or a logic analyzer. A logic probe can be used for quick testing, whereas for more in-depth testing more sophisticated test equipment is needed. It should be remembered that it is not suitable for many high speed logic circuits. It typically will only be useful for basic tests on basic circuits.

References

- Ohmmeter-circuit-and-types-of-ohmmeters: edgefx.in, Retrieved 31 March 2018
- Measuring-resistance-with-ohmmeter: engineering.electrical-equipment.org, Retrieved 11 April 2018
- How-to-use-a-multimeter, science-fair-projects: sciencebuddies.org, Retrieved 11 July 2018
- Spectrum-analyzer: searchnetworking.techtarget.com, Retrieved 28 May 2018
- What-is-a-logic-probe, test-methods: electronics-notes.com, Retrieved 11 March 2018

6

Applications of Electronics

Electronics is a vast subject that has a wide range of applications in areas such as audio electronics, radio electronics, nanoelectronics, molecular electronics and bioelectronics, among others. The diverse applications of electronics in the modern day have been thoroughly discussed in this chapter.

Audio Electronics

Audio electronics, which convert sound signals into electrical signals, process these electrical signals, and convert them back into sound.

The Basics of Sound

Before discussing the specific components of audio-related circuits and electronics, let's first take a crash course on the basic concepts of sound.

Sound is simply a type of energy vibrating through a medium (such as air or water); this energy, within a specific range of frequencies, is interpreted by the human ear as sound.

Sound is made up of three basic elements:

- Frequency: how fast the vibrations are occurring;
- Intensity: how loud the sound is;
- Timbre: the sound's quality.

The human ear can detect sound *frequencies* ranging from 20 to 20,000 Hz. However, the human ear is more sensitive to (i.e. able to discern at lowest intensity) frequencies ranging between 2,000 and 5,000 Hz. Recall that Hertz (Hz) is a unit defined as cycles per second.

A sound's *intensity* corresponds to the amount of energy associated with that sound. The decibel (dB) is used for measuring the sound's energy in a way that is relevant to how humans perceive loudness.

In the context of audio, the decibel is defined as follows:

$$dB = 10 \log_{10} (I/I_0)$$

where,

- I = the measured intensity (W/m^2);
- $I_0 = 10^{-12}$ W/m^2, which represents the lowest sound intensity detectable by the human ear.

The figure below shows a variety of sounds and their dB intensity measurement:

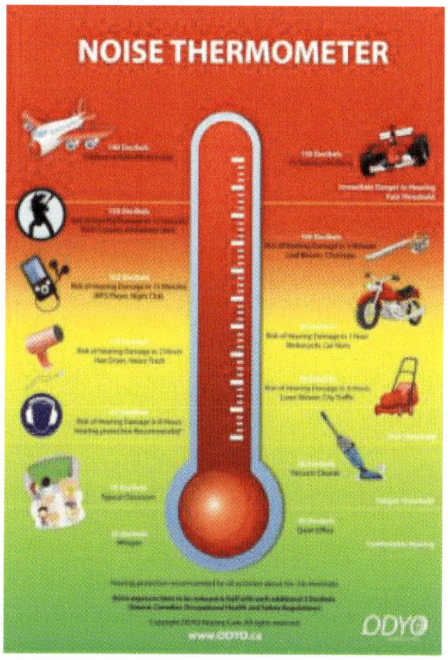

Examples of sounds and their intensity (in dB).

Timbre is the complex wave pattern that occurs when overtones (also referred to as harmonics) are present along with the fundamental frequency. For instance, a tuning fork is designed to produce exactly one specific frequency, called the fundamental frequency. Let's say we have a tuning fork that produces a middle C note (261.62 Hz. Since a tuning fork generates only one frequency, no other frequencies (overtones) exist.

However, if you were to hear a middle C played on a violin you would hear both the fundamental frequency (261.62 Hz) and a variety of other, albeit less intense, frequencies (overtones). The specific intensity of each overtone is largely credited for giving an instrument (or a person's voice) its unique timbre (also called tonal quality).

Microphones: Dynamic, Condenser and Electret

The microphone is responsible for converting changes in sound pressure to changes in electric current. The *intensity* of the changes in sound pressure corresponds to the AC voltage *amplitude* generated by the microphone. Likewise, the frequency of the changes in sound pressure corresponds to the frequency of the AC voltage. Obviously, if any overtones are present, they too are present in the electrical signal generated by the microphone.

Three common types of microphones are the following:

- Dynamic
- Condenser
- Electret

The Dynamic Microphone

The dynamic microphone consists of a plastic diaphragm, a voice coil and a permanent magnet.

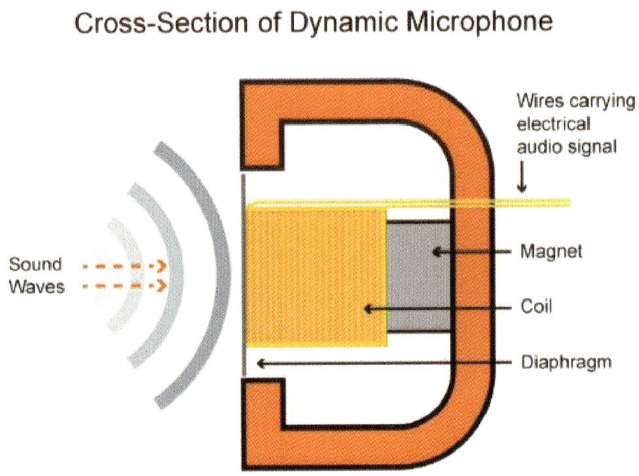

A dynamic microphone cross-section.

When varying sound pressure is applied to the diaphragm, the voice coil moves back and forth through the magnet's magnetic field, resulting in a voltage across the leads of the voice coil. Hence, the sound pressure is converted to a voltage signal.

Dynamic microphones are known for being very rugged, for being able to operate over a wide temperature range, for providing smooth and extended frequency response, and for not needing an external power source. They are widely used in applications such as public address and high-quality recording.

The Condenser Microphone

The condenser microphone uses a pair of charged plates that can be brought closer together or forced farther apart by changes in air pressure (i.e., sound). In doing so, the charged plates act like a sound-sensitive capacitor. This type of microphone works in conjunction with a low-noise, high-impedance amplifier.

Condenser microphones are known for providing crisp, low-noise sound and are used for producing high-quality sound recordings.

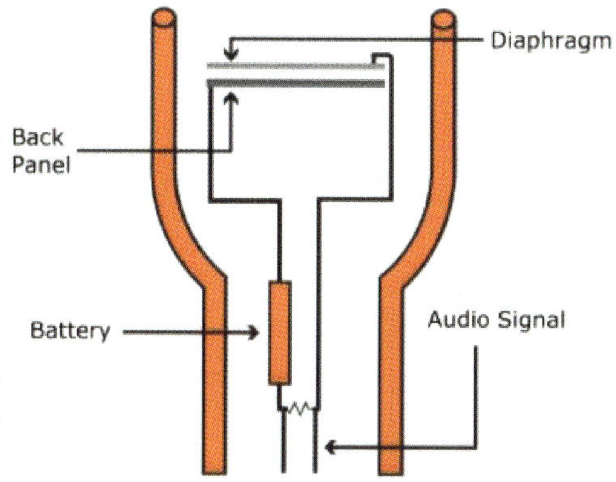

A condenser microphone's basic components.

The Electret Microphone

An *electret* microphone is actually a microphone and a JFET in a single package. The gate of the JFET provides a very high impedance and thus ensures that the charge on the electret element remains fixed. The charge must remain fixed to ensure that changes in capacitance (caused by sound waves) result in voltage changes. The JFET also converts the varying (gate) voltage into varying (drain-source) current. While the electret element itself does not require power (because its permanently charged) the overall electret microphone device requires a power supply because the integrated JFET needs bias current.

Older electret microphones were known to deliver poor performance, but modern devices can compete with condenser microphones.

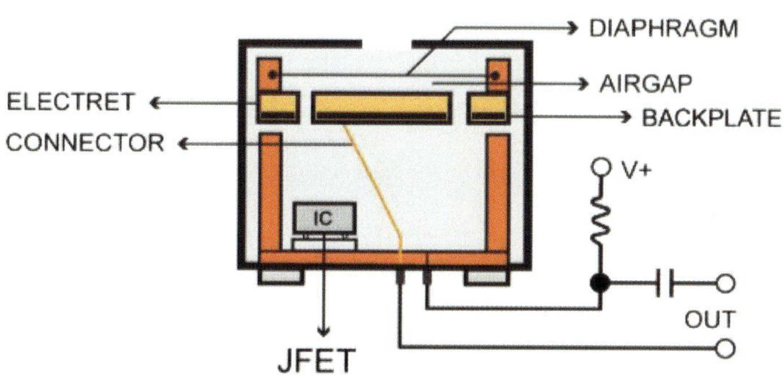

An electret microphone

Types of Speakers

Microphones convert sound signals into electrical signals. Speakers do the exact opposite they convert electrical signals into audible (sound) signals.

The dynamic speaker, which uses the same operating principles as the dynamic microphone, is the most popular speaker used today. When a varying electrical signal (current) is channeled through a moveable coil (voice coil) surrounded by a magnet, the coil moves back and forth. Cones (often made of paper) attached to the moving coil then produce variations in air pressure (i.e. sound waves) that correspond to the electrical signal driven through the speaker coil.

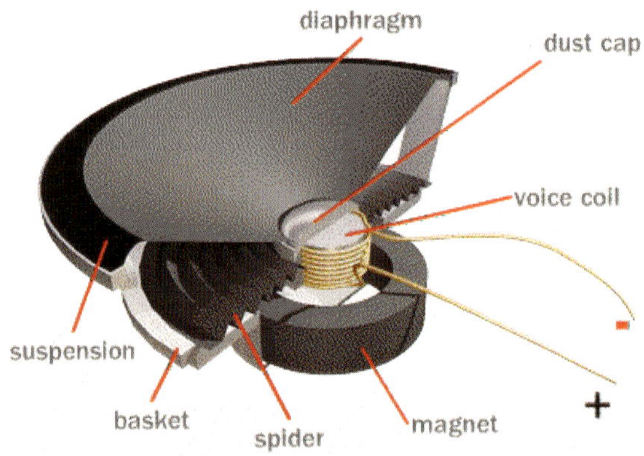

Parts of a dynamic speaker

Every speaker has a nominal resistance—or, more accurately, an impedance—that represents the average resistance between the speaker leads. Thus, speakers behave similarly to resistors when placed in series and in parallel.

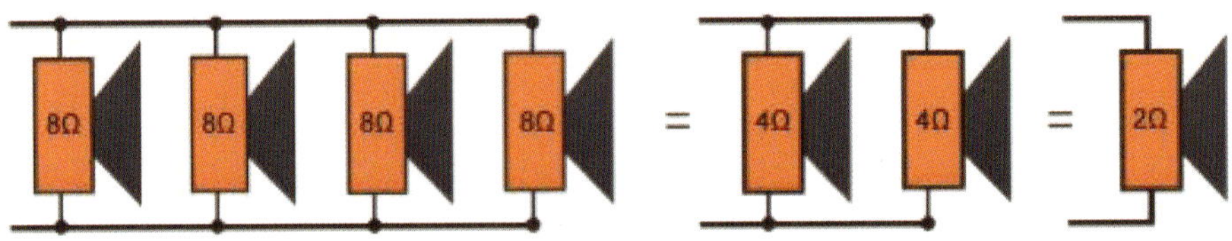

Speakers in parallel behave the same as resistors in parallel.

Frequency response, an important speaker characteristic, represents the frequency range over which a speaker can successfully propagate sound waves.

Speakers can be categorized according to the frequency range that they are designed for:

- Woofers: speakers designed specifically for low frequencies (less than 200 Hz);
- Midrange: speakers designed to accommodate frequencies ranging from 500 Hz to 3000 Hz;
- Tweeter: a dedicated speaker type specifically designed to handle frequencies above those of midrange speakers;
- Full-range speakers: capable of handling frequencies ranging from 100 Hz to 15,000 Hz.

Full-range speakers have inferior sound qualities when compared to a sound system consisting of woofers, midrange speakers, and tweeters. Such a three-way speaker system is illustrated below:

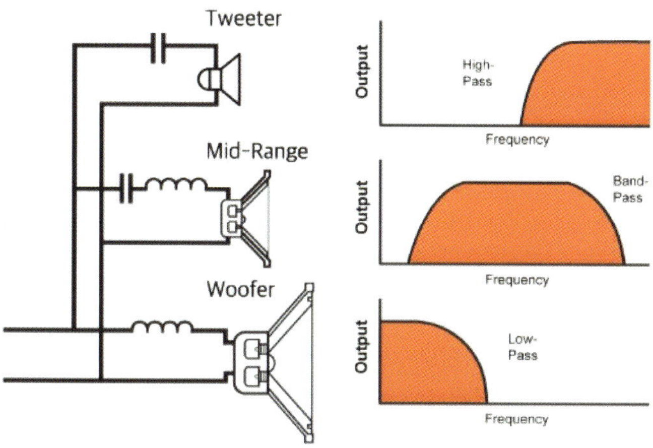

A typical three-way speaker system consisting of a tweeter, a midrange speaker, and a woofer

Amplifiers 101

Audio amplifier have three types of classification:

- Preamplifier
- Low-power amplifier
- Power amplifier

An audio preamplifier (often shortened to "pre-amp") is an electronic device that amplifies a very weak signal from a microphone, as an example, into signals strong enough to manipulate. Pre-amps are often simple, fixed-gain amplifiers designed specifically for low-noise performance.

A *low-power amplifier* is often used to manipulate signals including such aspects as volume and frequency equalization. This type of amplifier generally focuses on changing the character of the signal in desired ways while introducing as little unwanted distortion as possible and may provide little to no actual power amplification.

An audio *power amplifier* ("power amp") is used to increase the signal power so as to drive a load, such as output speakers. Similar to pre-amps, power amps are often fixed gain (in terms of signal amplitude) so that designers can focus on high-power gain and the power handling challenges that typically result. In simple audio systems where high power and high fidelity are not critical factors, a single-amplifier circuit may perform all of these functions, and, in fact, specially designed operational amplifiers, such as the LM-386 Low Voltage Audio Power Amplifier, are often used this way. A Class D audio amplifier is a switching or PWM amplifier where the switches are either fully ON or fully OFF, which significantly reduces the power losses. Class D amplifiers take on many different forms, including some that have digital inputs while others have analog inputs.

Radio Electronics

Radio Electronics is a term encompassing a wide range of areas in science and engineering that deal with the problems of transmission, reception, and processing of information by means of electromagnetic waves. The term came into use during the 1950's and to a certain extent is conventional.

Radio electronics encompasses radio engineering and electronics, as well as a number of new areas that have emerged as a result of development and differentiation. Such areas include quantum electronics, optoelectronics, semiconductor electronics, microelectronics, infrared technology, cryoelectronics, acoustical electronics, and chemotronics. Radio-electronics is closely associated with radio physics, solid-state physics, optics, and mechanics, as well as with electrical engineering, automation, and engineering cybernetics. Radio-electronic apparatus often serve as one of the parts inautomatic control systems, such as guidance systems that govern the flight of rockets and space craft. Conversely, automatic control systems may be used in such apparatus, for example, in automatic frequency control or target tracking.

Radio electronics is also associated with computer engineering, since this branch of engineering uses electronic devices inthe processing of information, as in noise suppression or reduction of data to a pre-determined form. Frequency bands that are used in radio electronics overlap those used in electroacoustics. Mathematical studies are widely used in radioelectronics for the analysis and synthesis (design) of radio-engineering circuits and devices and for the determination ofoptimum design and parameters.

Consumer Electronics

Consumer electronics includes a broad set of electronic devices that provide one or more functionalities in a home or for a particular individual. The term initially referred to electronic devices that were installed or used specifically inside a home/house. However, they also now incorporate mobile and computing devices, which can easily be carried by an individual outside the home, such as a cell phone or a tablet PC.

Consumer electronics include items, such as:

- Televisions

- DVD players
- Refrigerators
- Washing machines
- Computers
- Laptops
- Tablets

Nanoelectronics

Nanoelectronics is based on the application of nanotechnology in the field of electronics and electronic components. Although the term Nanoelectronics may generally mean all the electronic components, special attention is given in the case of transistors. These transistors have a size lesser than 100 nanometres. Visibly, they are very small that separate studies have to be made for knowing the quantum mechanical properties and inter-atomic design. As a result, though the transistors appear in the nanometre range, they are designed through nanotechnology. Their design is also very much different from the traditional transistors and usually falls in the category of one-dimensional nanotubes/nanowires, hybrid molecular electronics, or advanced molecular electronics.

This technology is said to be the next future, but its practicality is near to impossible even now that they may be difficult to emerge soon.

Basic Concept of Nanoelectronics

Although a nanoelectronic device can be made fully functional, the work load it can do is restricted to its size. The basic principle is that the power of a machine will increase according to the increase in volume, but the amount of friction that the machine's bearings hold will depend on the surface area of the machine.

For the small size of the nanoelectronic device cannot be used for the moving of heavy load like a mechanical device. If such a task is tried, it will fail as the available power will be easily overcome by the frictional forces. So, it is sure that these devices have limitations in real world applications.

Different Approaches to Nanoelectronics

- Nanofabrication

This method is used to design arrays or layers of nanoelectronic device to work for a single operation. Nanoelectro-mechanical systems are also a part of nanofabrication.

- Nanomaterials electronics

In Nanoelectronics, the transistors are packed as arrays on to a single chip. Thus they remain in a uniform manner and symmetrical in nature. Thus they are known to have a more speedy movement of electrons in the material. The dielectric constant of the device also increases and the electron or hole characteristics also become symmetrical in nature.

Besides transistors, nanoelectronic devices play a role in data storage (memory). Here, *spintronics* – the study and exploitation in solid-state devices of electron spin and its associated magnetic moment, along with electric charge is already an established technology.

Illustration of electron spin in a graphene lattice.

Spintronics also plays a role in new technologies that exploit quantum behavior for computing.

Applications of Nanoelectronics

Some of the devices that have been developed with the help of Nanoelectronics and its future applications are listed below:

- Nanoradio
- Nanocomputers

The conventional computers with a big processor will be replaced with Nanocomputers with nanoprocessors that will have higher performance and speed than the conventional computers. Researchers are performing various experiments on by using nanolithographic methods to design better nanoprocessors. Experiments are also taking place by replacing the CMOS components in conventional processors with nanowires. The FET's in the computers are replaced by carbon nanotubes.

- Energy production

The devices using Nanoelectronics technology also includes solar cells that are highly efficient and cheaper than the conventional ones. If such efficient solar energy can be created it would be a revolution to the global energy needs.

Using the technology, researchers are developing a generator for energy production in vivo called bio-nano generators. Basically, the generator is an electrochemical device which is designed in nanoscale size. It works like a fuel cell which generates the power by absorbing the blood glucose in a living body. The glucose will be separated from the body with the help of an enzyme. This enzyme separates the glucose from the electrons and makes them useful for generating power.

The power generated through such a device will be only a few watts as the body itself needs some glucose for its normal functioning. This small power can be used to power up devices placed inside the body like pacemakers or sugar-fed nanorobots.

Molecular Electronics

Molecular electronics is an interdisciplinary field that involves the use of molecules as fundamental electronic components such as wires, transistors, memory cells, and logic elements. The field, sometimes called "moletronics" is currently attracting attention from a broad cross-section of the scientific community in response to both fascination with the fundamental scientific challenges associated with the measurement of the electrical properties of molecules, and manipulation of electronic phenomena via molecular processes, and growing concerns over the technological challenges and ultimate limits facing solid state semiconductor technology.

Before entering discussion, a distinction needs to be made between molecular materials for electronics and single molecule electronics. The use of organic materials for electronic applications in which the bulk electronic or optoelectronic response arises from ensembles of several millions of molecules, and for which properties are measured or observed on the macroscopic level, is at a mature stage of development and application. The most obvious examples of molecular materials readily available in the electronics mass market are the use of liquid crystals and organic light emitting diodes (OLEDs) in flat video displays. This sector of the electronics industry continues its steady development driven by the promises of transparent, flexible, non-toxic, printable electronics, with improved consumer appeal and reduced fabrication costs.

However, the field of single molecule electronics, which is the subject of this review, can be summarised as the use of individual molecules to mimic functional elements in electronic devices. The use of molecules as building blocks to give rise to a more complex system is typically referred to as a the "bottom-up" approach, as opposed to the conventional "top-down" lithographic techniques that are employed to etch small features into single-crystal silicon wafers. Despite, or perhaps because of, the great challenges the field of single molecule electronics presents, the enormous benefits of a molecule-based "bottom-up" approach motivates the scientific community to keep moving the field forward. For example, a typical computer microchip nowadays contains of the order of

109 transistors in areas of about 3 cm 2 with all its components working in perfect harmony and capable of changing state in response to an input signal at an astonishing rate and for an almost unimaginable number of cycles. The enormous challenge for single molecule electronics lies in the development of molecular systems that perform in some way better than, or perhaps provide an alternative function to, these extraordinary solid-state devices.

Bioelectronics

Bioelectronics is the application of the principles of electronics to biology and medicine. Hollywood has expanded the horizons with its imagination, and movies like Star Wars, Star Trek, iRobot, Minority Report and many other sci-fi movies have shown us what modern technology is capable of. The potential of bioelectronics is also shown in these movies. Artificial limbs, humanoids, the various sensors that are attached to the body, etc. are the applications of bioelectronics.

Bioelectronics encompasses a range of topics at the interface of biology and electronics. One aspect of bioelectronics is the application of electronics to problems in biology, medicine, and security. This includes electronics for both detection and characterization of biological materials, such as on the cellular and subcellular level. Another aspect of bioelectronics is using biological systems in electronic applications (e.g., processing novel electronic components from DNA, nerves, or cells). Bioelectronics also focuses on physically interfacing electronic devices with biological systems (e.g., brain-machine, cell-electrode, or protein-electrode).

Applications in this area include assistive technologies for individuals with brain-related disease or injury, such as paralysis, artificial retinas, and new technologies for protein structure-function measurements.

Examples of Technical Areas of Opportunity for Bioelectronics

The overarching technical drivers for bioelectronics are the constant advances in semiconductor technology coupled with advances in surface chemistry and their application to life sciences research. Much of the work that has been done in the bioelectronics field since the 1990s has been focused on creating better biosensors by integrating biomolecules with semiconductors. For decades, semiconductor technology has advanced at an exponential rate and is described by Moore's law, which states that the number of features in a given area of substrate doubles every 18–24 months. A result of Moore's law is that the computing power and capabilities increase with each generation of device while the cost per function decreases.

As Moore's law has progressed, so have the number of semiconductor applications in the life sciences. In particular, significant effort has gone into developing surface chemistries that can be used to attach biological molecules to semiconductor substrates. Two examples are biosensors that use enzymes covalently linked to electrode surfaces and DNA recognition based on surface-bound DNAfunctionalized polypyrrole molecules. Although advances have been made in binding biological molecules to substrates, no truly biomimetic synthetic surfaces yet exist (e.g., for use in implants).

The following are examples of technical areas or opportunities that will impact bioelectronics:

Real-Time and Massively Parallel Molecular and Cellular Characterization for Systems Biology

The nascent field of systems biology using systems engineering approaches to analyze cellular function is driving the development of new technology that can monitor multiple aspects of cellular behavior over many timepoints. Systems biology embodies a new perspective from which to view biological systems and knowledge culled from its approaches could lead to advances in medicine and security. However, significant investment is needed to develop tools and associated standards and metrology that can characterize and continuously monitor the states of cells at sub-cellular resolutions. Advances in micro- and nano-fluidic systems that can monitor many cell populations in parallel are beginning to address this need.

Biologically-based Sensors and Fabrication

Another driver of bioelectronics is that cells and their components can be used as biological transducers for measurements or as components in building novel materials or circuits. In other words, components of cells have been used for novel applications outside of their originally observed purpose. Cells are sensitive to many environmental cues. For example, identifying the response of a cell to a known toxin could allow it to be used as a "canary in a coal mine" to detect toxic substances in captured air samples. Biomolecules, in particular antibodies can also be used as transducers, via their exquisite specificity for complementary molecules. Coupling antibodies with emerging nano-scale technologies could result in ultra-sensitive detection methods. Bio-inspired fabrication shows promise for constructing nanoscale assemblies, which could lead to significant advances in sensor design and materials technology.

Protection and Restoration of Health

The development of new technology that can protect lives (e.g., by preventing bioterrorism) and restore health (e.g., by developing novel implantable therapeutic devices) is also driving the development of bioelectronics. National security demands new technology for monitoring chemical or biological threats, and advances in lab-on-a-chip technology have been developed to address this need. Further miniaturization of these technologies will likely result in smaller units with increasing capabilities. Advances in miniaturization and power transmission, storage, and generation have allowed implantable medical devices to emerge. The artificial retina, which is a first step toward restoring sight in people with degenerative diseases of the retina, is representative of these advancements. Implantable drug delivery devices, based on micro-electro-mechanical systems (MEMS), are beginning to emerge, and the next generation of drug delivery devices are likely to be "smart" and to have embedded transceivers (i.e., able to sense and respond to their environment and transmit and receive data) instead of passively delivering drugs at pre-defined intervals.

Applications of Bioelectronics

Pacemakers: Also called artificial pacemakers, have been a boon to patients by regulating their heart beats. Initially they weren't monitored electronically but eventually due to the development in the field of bioelectronics and biotechnology they work perfectly, have given life to thousands across the world, and make complex problems concerned with the heart simple.

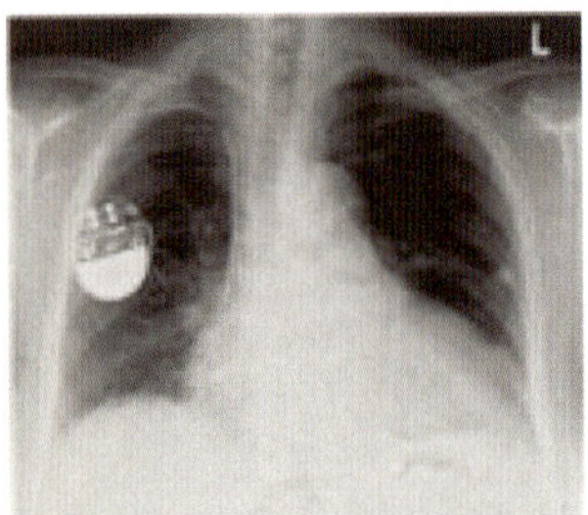

Artificial Limbs: For people who have lost their limbs in accidents or some natural cause prosthetics are used. Bioelectronics is used to power them, control them, modify and manipulate their structure as per the needs. Thus the affected ones are given a second chance to live thanks to bioelectronics and its principles.

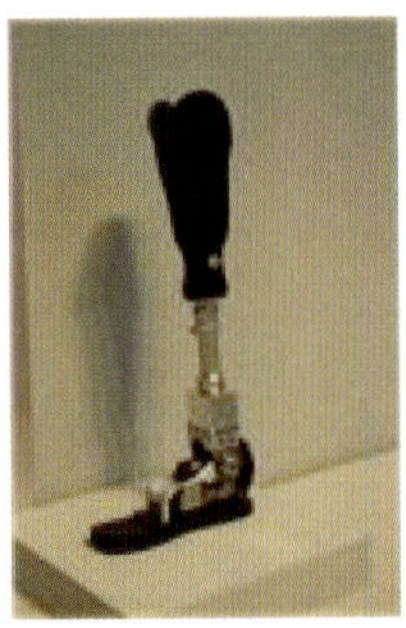

Blood Glucose Meter: Patients with diabetes need to keep their blood glucose level in check. Various blood glucose meters are available in the market that help patients to check their blood glucose in their own home instead of the olden days method of going to some centers and getting the blood glucose level checked. They give accurate readings with just a couple of drops of blood and there is virtually no pain in the entire process. Thus the patient can keep check of his or her blood glucose level and this will keep the patient out of danger.

Nuclear Electronics

Nuclear Electronics is the aggregate of methods of nuclear physics in which electronic instruments are used to receive, convert, and process information from nuclear radiation detectors. In addition to applications in nuclear physics and elementary particle physics,such methods are used wherever work involves ionizing radiation—in chemistry, medicine, space research and elsewhere.

Because of the short duration and (usually) high frequency of the processes in question as well as the presence ofbackground noise, instruments used in nuclear electronics must have a high time resolution (~10^{-9} sec). Because of the necessity of simultaneous measurement of a large number of parameters (signal amplitude and arrival time, the coordinates of the detection point, and so on), nuclear electronics was the first field for which analogue-digital

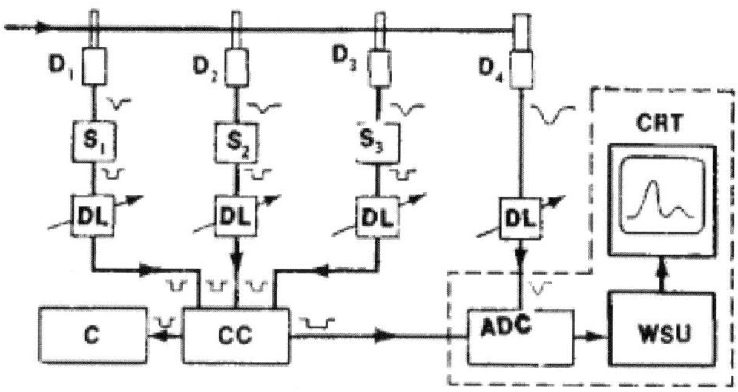

Figure: Diagram of a charged-particle spectrometer: (D_1), (D_2), (D_3), and (D_4) detectors, (S_1), (S_2), and (S_3) pulse shapers, (CC)coincidence circuit, (DL) delay lines, (ADC) analogue-to-digital converter, (WSU) working storage unit, (CRT) cathode-ray tube, (C) counter

Conversion circuits were developed, digital methods of data accumulation were used, multichannel and multidimensionalanalyses were conducted, and computers were used.

In the registration of particles (or quanta) the task of nuclear electronics is to count pulses received from a detector. In the identification of the type of radiation or the investigation of the radiation's spectrum, analysis is made of pulse wave form, pulse amplitude, or the relative delay between pulses. In the investigation of the spatial distribution of radiation, the numberof detectors that have fired is recorded or the coordinate of the detection point may be determined directly.

The principal elements of the devices used in nuclear electronics include coincidence circuits, anti-coincidence circuits, amplitude discriminators, linear gating circuits and adders, multichannel time and amplitude analyzers and various devices that give a plot of information from coordinate detectors such as spark chambers and proportional ionization chambers. A complete listing would include hundreds of items.

A device for registering particles contains a detector, amplifier, signal converter and registering device. The converter transforms the detector signal into a standard pulse or converts the amplitude or arrival time of the signal to digital code. Pulse counters, memory devices or computers may be used to record measurement results; chart recorders and camerasare less often used.

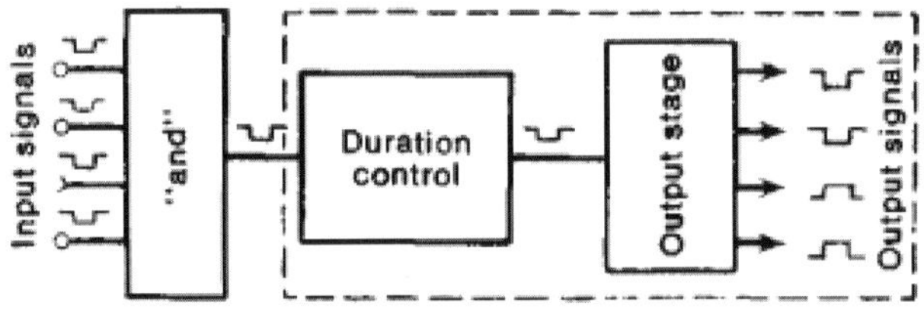

Figure: Coincidence circuit

A simplified system for studying radiation spectra is shown in figure above. A charged particle intersects detectors (D_1), (D_2) and (D_3) and is stopped in detector (D_4). The signals from detectors (D_1), (D_2) and (D_3) are fed through pulse shapers (S_1), (S_2) and (S_3) to a coincidence circuit (CC), which selects events whose signals arrive at its inputs simultaneously. Simultaneity of pulse arrival is ensured by delay lines (DL). The coincidence circuit produces a signal that allows conversion of the pulse under investigation from detector (D_4). The converted pulse is fed as a digital code from an analogue-to-digital converter (ADC) to a working storage unit (WSU) or a computer. The measured amplitude spectrum is displayed on the screen of acathode-ray tube (CRT). This part of the system, enclosed within the broken line, is a multi-channel amplitude analyzer. The count rate at the output of the coincidence circuit, as read out by a counter (C), shows the number of registered events. The time selection of signals is handled by coincidence circuits that operate from pulses with a specified duration and amplitude. The coincidence circuits execute the logical operation "and" (logical multiplication); that is a signal appears at the output ofthe circuit only when the pulses at all inputs are of a specified level, called the unit level. If a signal with inverted polarity isapplied to one of the inputs of a coincidence circuit, the circuit becomes an anti-coincidence circuit. Modern coincidence and anti-coincidence circuits use standard integrated circuitry.

Amplitude selection is handled by discriminators based on a Schmitt trigger or tunnel diodes. The discriminators produce a standard output pulse only if the voltage (or current) at the input exceeds a specified level. Comparators are often used for amplitude discrimination. The evolution of coincidence circuits and amplitude discriminators is typical of other nuclearelectronics instruments as well. In addition to units that execute a single logical operation ("and," "or" and so on), universal multi-operation devices have been developed in which the logical operation can be assigned externally. The development of such devices has been made possible by the introduction of computers in nuclear electronics.

Computer technology has also facilitated the development of automated equipment with program control of parameters: the computer controls the operating thresholds of gating circuits, time resolution, signal delay, the logic of event selection, and the mode of operation of the measuring system. Microprocessors and specialized processors for pattern recognition and for accumulation and preliminary processing of measurement results have also been introduced into practice in physicale xperiments. Experimental data are accumulated in the computer with subsequent transcription on to magnetic tape. The results of preliminary processing are displayed on the screen of a cathode-ray tube, enabling the operator to intervene during the measurement process. The computer controls various actuating devices motors that move the detectors or target, relays, signal commutators and the like.

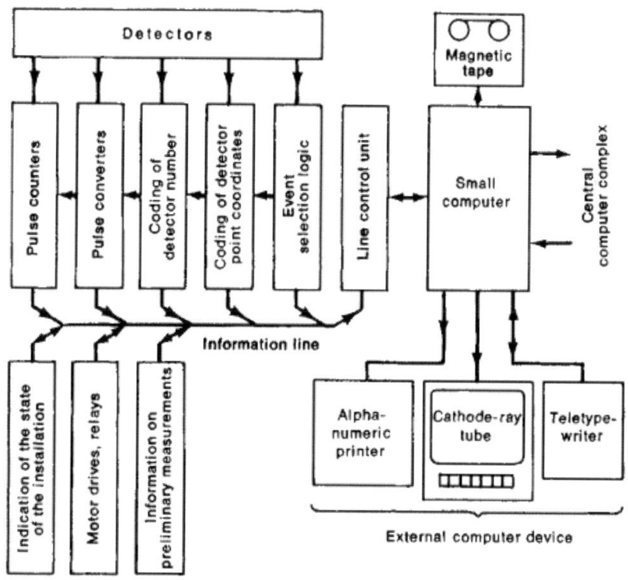

Figure: System for accumulating and processing information in a nuclear physicsexperiment

References

- Introduction-audio-electronics-sound-microphones-speakers-amplifiers: allaboutcircuits.com, Retrieved 05 March 2018
- Consumer-electronics-ce-757: techopedia.com, Retrieved 25 April 2018
- Nanoelectronics: circuitstoday.com, Retrieved 05 July 2018
- Molecular-Electronics-Revised-11601647: researchrepository.uwa.edu.au, Retrieved 05 April 2018
- Overcoming-disability-with-bioelectronics-73831, consumer-appliances-electronics: brighthubengineering.com, Retrieved 25 June 2018

Permissions

All chapters in this book are published with permission under the Creative Commons Attribution Share Alike License or equivalent. Every chapter published in this book has been scrutinized by our experts. Their significance has been extensively debated. The topics covered herein carry significant information for a comprehensive understanding. They may even be implemented as practical applications or may be referred to as a beginning point for further studies.

We would like to thank the editorial team for lending their expertise to make the book truly unique. They have played a crucial role in the development of this book. Without their invaluable contributions this book wouldn't have been possible. They have made vital efforts to compile up to date information on the varied aspects of this subject to make this book a valuable addition to the collection of many professionals and students.

This book was conceptualized with the vision of imparting up-to-date and integrated information in this field. To ensure the same, a matchless editorial board was set up. Every individual on the board went through rigorous rounds of assessment to prove their worth. After which they invested a large part of their time researching and compiling the most relevant data for our readers.

The editorial board has been involved in producing this book since its inception. They have spent rigorous hours researching and exploring the diverse topics which have resulted in the successful publishing of this book. They have passed on their knowledge of decades through this book. To expedite this challenging task, the publisher supported the team at every step. A small team of assistant editors was also appointed to further simplify the editing procedure and attain best results for the readers.

Apart from the editorial board, the designing team has also invested a significant amount of their time in understanding the subject and creating the most relevant covers. They scrutinized every image to scout for the most suitable representation of the subject and create an appropriate cover for the book.

The publishing team has been an ardent support to the editorial, designing and production team. Their endless efforts to recruit the best for this project, has resulted in the accomplishment of this book. They are a veteran in the field of academics and their pool of knowledge is as vast as their experience in printing. Their expertise and guidance has proved useful at every step. Their uncompromising quality standards have made this book an exceptional effort. Their encouragement from time to time has been an inspiration for everyone.

The publisher and the editorial board hope that this book will prove to be a valuable piece of knowledge for students, practitioners and scholars across the globe.

Index